<u>Disclaimer</u>

Book Title: Measurement of Temperature, Displacement, and Strain in Structural Components Subject to Fire Effects: Concepts and Candidate Approaches

Book Author: Therese P. McAllister; William E. Luecke; Mark A. Iadicola; Matthew F. Bundy;

Book Abstract: For the last forty years, NIST has led the world in fire metrology through research conducted at the Large Fire Laboratory, which is being expanded to enable experiments on real-scale structures under combined structural and fire loads. The combined capabilities of large fire testing and structural fire testing will be comprised in the National Fire Research Laboratory (NFRL), which is expected to be completed in 2013. Measurements of temperature, displacement, and strain at hundreds of points on a structural system in the fire zone are needed to validate analytical tools for fire conditions. However, the ability to measure the performance of structures during realistic fire exposures is severely limited due to a significant gap in measurement science. At present, temperatures are measured with thermocouples and strains are measured with high temperature strain gages. Each of these sensors requires a separate line for data collection during the experiment. Further, high temperature strain gages are unreliable and often do not perform as expected during fire tests. Significant improvements to structural measurement in fire conditions are needed to advance the validation of analytical tools and performance based design methodologies. Candidate methods for temperature, displacement, and strain measurements that could meet these performance requirements were reviewed. A demonstration test that employed a natural gas burner in the Large Fire Facility evaluated the potential of digital image correlation and high temperature strain gages to measure thermally induced strains.The technology review and the outcome of the demonstration test indicate that digital image correlation and fiber optic methods have great promise for temperature, displacement, and strain measurement. A four-stage development plan is proposed to overcome these challenges.

Citation: NIST TN - 1768

Keywords: structural fire tests; strain measurement; displacement measurement; temperature measurement; fiber optic; digital image correlation

NIST Technical Note 1768

Measurement of Temperature, Displacement, and Strain in Structural Components Subject to Fire Effects: Concepts and Candidate Approaches

Therese McAllister
William Luecke
Mark Iadicola
Matt Bundy

http://dx.doi.org/10.6028/NIST.TN.1768

National Institute of
Standards and Technology
U.S. Department of Commerce

NIST Technical Note 1768

Measurement of Temperature, Displacement, and Strain in Structural Components Subject to Fire Effects: Concepts and Candidate Approaches

Therese McAllister
Materials and Structural Systems Division
Engineering Laboratory

William Luecke
Mark Iadicola
Metallurgy Division
Material Measurement Laboratory

Matt Bundy
Fire Research Division
Engineering Laboratory

http://dx.doi.org/10.6028/NIST.TN.1768

November 2012

U.S. Department of Commerce
Rebecca Blank, Acting Secretary

National Institute of Standards and Technology
Patrick D. Gallagher, Under Secretary of Commerce for Standards and Technology and Director

Certain commercial entities, equipment, or materials may be identified in this document in order to describe an experimental procedure or concept adequately. Such identification is not intended to imply recommendation or endorsement by the National Institute of Standards and Technology, nor is it intended to imply that the entities, materials, or equipment are necessarily the best available for the purpose.

National Institute of Standards and Technology Technical Note 1768
Natl. Inst. Stand. Technol. Technical Note 1768, 73 pages (November 2012)
http://dx.doi.org/10.6028/NIST.TN.1768
CODEN: NTNOEF

Executive Summary

For the last forty years, NIST has led the world in fire metrology through research conducted at the Large Fire Laboratory, which is being expanded to enable experiments on real-scale structures under combined structural and fire loads. These two capabilities will be joined in the National Fire Research Laboratory (NFRL), which is expected to be completed in 2013. Measurements of temperature, displacement, and strain at hundreds of points on a structural system in the fire zone are needed to validate analytical tools for structural response in fire conditions. However, the ability to measure the performance of structures during realistic fire exposures is severely limited due to a significant gap in measurement science.

At present, temperatures are measured with thermocouples and strains are measured with high-temperature strain gages. Each of these sensors requires a separate signal path for data collection during the experiment. Further, high-temperature strain gages are unreliable and often do not perform as expected during fire tests. Significant improvements to structural measurement in fire conditions are needed to advance the validation of analytical tools and performance based design methodologies.

A set of desired performance criteria for structural-fire measurements were established to guide the assessment and development of candidate technologies and measurement systems. Rather than having one measurement per sensor/data line, measurements over an entire surface or multiple sensors per data line are desired. The ability to record tens to hundreds of data points per line will produce the quantity and quality of data needed for understanding structural behavior and validating analytical tools. Measurements are necessary at temperatures as high as 750 °C, and they must be insensitive to heating rate. Displacement measurements can range from 1 mm to 1 m. Elastic strain measurements up to $e = 0.002$ with a resolution of $\pm 200 \ \mu m/m$ and plastic strain measurements up to $e = 0.2$, with a resolution better than $\pm 2000 \ \mu m/m$ are the desired measurement capabilities.

Candidate methods that could meet these performance requirements were reviewed. While technologies that could provide multiple measurements at elevated temperatures were the primary focus, technologies that could provide single-point measurements at elevated temperatures were also considered.

- Candidate methods for temperature measurement include thermocouples, infrared imaging sensors, and fiber-optic sensors.

- Candidate methods for displacement measurement include linear transducers, laser distance measurement, and digital image correlation techniques.

- Candidate methods for strain measurement include traditional high-temperature strain gages, digital image correlation techniques, and fiber-optic sensors.

A demonstration test that employed a natural gas burner in the Large Fire Facility evaluated the potential of digital image correlation and high-temperature strain gages to measure thermally induced strains. Digital image correlation successfully measured the resulting thermal expansion strain of a plate heated by fire; the strain gages did not provide meaningful data in fire conditions.

The technology review and the outcome of the demonstration test indicate that digital image correlation and fiber-optic methods have great promise for temperature, displacement, and strain measurement in the NFRL, but many challenges must be overcome. For digital image correlation, the challenges include ensuring the integrity of the patterned coating, forming usable images in spite of the optical radiation from the fire, processing to minimize image changes due to turbulence, and coping with smoke and soot deposition. For fiber-optic sensors challenges include preserving the integrity of the sensor at high temperature and ensuring that the strain in the member is transferred to the fiber. A method to deconvolve the contributions of thermal, elastic, and plastic strains is needed for both measurement technologies.

A four-stage development plan is proposed to overcome these challenges. Two early stages develop the measurement science with simple heating and deformations. The intermediate stage employs more realistic fires and more complicated deformations. In the final stage, the measurement uncertainty of both methods will be quantified.

If the proposed research is successful, the resulting structural metrology in fire will greatly enhance the research capabilities of the NFRL. By starting this new line of foundational research, structural measurements in the fire zone could be propelled from virtually nonexistent to full-field measurements.

Keywords: structural fire tests; strain measurement; displacement measurement; temperature measurement; fiber-optic; digital image correlation.

iii

Acknowledgements

Many colleagues, both inside and outside NIST contributed to this report. NIST colleagues include Francine Amon, Peter Bajcsy, Rodney Bryant, Stephen Cauffman, Artur Chernovsky, Ashu Gola, Sae Woo Nam, Long Phan, Kuldeep Prasad, Richard Rhorer, and Eric Whitenton. Colleagues outside of NIST include Pedro Calderon at the Polytechnic University of Valencia, Gustave Fralick at NASA, Otto Gregory at the University of Rhode Island, Alex Sang at Luna Technology, Anbo Wang at Virginia Tech University, and Vince Wnuk at HiTec Products.

Contents

v

1.   Introduction

Temperature, displacement, and strain measurements of structural components and systems are needed to fully understand the performance of structural systems in fire and to validate analytical models of structural response to fire effects. Present-day tools, such as the prescriptive ASTM E119 standard, provide comparative fire ratings between assemblies, but no useful measured data about structural performance. Measurements of structural performance in fire conditions are needed to validate analytical tools and to develop performance-based design methodologies.

Most structural fire resistance experiments are conducted either in furnaces or for compartment fire conditions. For most experiments, structural displacements and strains are not measured in the heated zone of a fire test, where transient gas temperatures may reach 1400 °C. Instead, structural displacements are measured outside the heated zone at unheated slab surfaces or at unheated column ends, and strains are not usually measured.

Measurement techniques for structural response at room temperature and with thermocouples for temperature are well developed, but measurement techniques for structural response in the fire zone at elevated temperatures are lacking. For example, Figure 1 shows a composite floor truss system constructed to replicate the World Trade Center (WTC) towers floor system that was tested according to the ASTM E119 standard. Heat flux, gas temperatures, steel surface temperatures, and vertical displacement at nine locations on the unheated side of the floor slab were measured; no local deformation or strain measurements were obtained. These state-of-the-art measurements are inadequate for understanding structural behavior in fire conditions and for validating analytical tools.

Measurements under realistic fire conditions require instrumentation that can withstand elevated temperatures and the evolving fire environment (for example varying states of smoke, convective air distortions, and soot), while providing accurate, reliable, and repeatable measurements. Additionally, strain measurements need to be resolved into thermal and mechanical components. Total strain measurements may comprise thermal strain, elastic strain, plastic strain, and creep strain, each of which depends on temperature.

Room-temperature strain and displacement measurement devices primarily measure response at discrete points which must be selected *a priori*, based on either experience or a predictive analysis model. The discrete measurement devices are located at select locations, and may miss critical data, reducing the value of expensive, complex laboratory experiments and potentially result in erroneous conclusions. The likelihood of missing critical locations during fire tests is even greater, due to thermally induced forces that evolve during the fire exposure. In such situations, field measurement systems that record structural response data across surfaces (for example a camera-based measurement system) or along/through components (for example a fiber-based measurement system) would provide a much richer data set. Such systems have been developed for room-temperature applications and would provide a tremendous leap in structural-fire measurement capabilities.

Figure 1. Scaled model of the composite floor system in the WTC Towers during an ASTM E119 test. Note deformations in truss web members, which were observed but not measured.

The ability to measure displacement, strain, and temperature across structural components during fire exposure will:

- provide comprehensive experimental data on the structural performance of components and connections in systems under realistic fire conditions and gravity loads,

- support the validation of physics-based analytical tools,

3

- provide the technical basis for the development of performance-based standards for the design of fire resistant structures,

- foster innovation in design and construction, and

- lead to the development of improved building materials, standards, design practice, and building codes.

The specific challenges of developing structural response measurements for fire conditions are primarily to develop, through adaptation or innovation, field measurement technologies for ambient applications to the harsh environment of fire and to quantify measurement resolution and uncertainty as a function of temperature.

The NIST National Fire Research Laboratory (NFRL) will be available for conducting full-scale structural-fire tests with realistic fire and boundary conditions in 2013. The utility of the facility and test data will be greatly increased with accurate measurements of the structural response (displacements and strains at known temperatures) in fire.

2. Measurement Methods in Structural-Fire Test Laboratories

The review of structural measurement methods under fire conditions included determining which, if any, methods are used at other structural fire test facilities. Table 1 lists the displacement and strain measurement methods used at Michigan State University, Purdue University, University of Coimbra, Portugal, Centre Technique Industriel de la Construction Metallique (CTICM), France, the National Research Council (NRC) of Canada, the British Research Establishment (BRE) Cardington Facility, the Architecture and Building Research Institute (ABRI) Fire Research Facility, Taiwan, and the Korean Institute of Construction Technology (KICT), Fire Research Center. This is a representative list of structural fire testing capabilities around the world. A recent NIST publication [1] summarizes the capabilities of other fire test facilities. All of the test facilities listed in Table 1 measure gas and structural temperatures with thermocouples inside the furnaces, as they must meet the standard fire heating requirements.

2.1. Michigan State University

The furnace housed at the Civil Infrastructure Laboratory at Michigan State University is capable of testing loaded structural assemblies such as columns, beams and floor systems under fire conditions [2]. It consists of a steel framework supported by four columns with the furnace chamber of about 2.5 m by 3 m. Two small view ports on either side of the furnace wall allow visual monitoring of the fire-exposed surface during tests.

Several measurement methods have been used for obtaining structural displacement and strain data [2,3,4]. Displacements are measured with Linear Variable Differential Transformer (LVDT) devices outside of the heated zone. Laser-based devices did not work in a fire environment due to fire spectrum interference at the laser frequency. Strain measurements of steel components were attempted with high-temperature strain gages. They failed in the first 20 minutes of an ASTM E119 fire exposure, when temperatures reached 400 °C to 600 °C. Bonding adhesion of the strain gages to steel surfaces was also a problem at elevated temperatures. The strain gages needed to be welded for temperatures up to 800 °C. Concrete strain gages worked until the concrete cracked, which released the local strain.

2.2. Purdue University

Purdue University researchers designed a system of heating panels to simulate the heating effects of fire. The panels have electrical coils, and are placed close to surfaces being tested. The heating system is being used to test full-scale steel columns at the Bowen Laboratory for Large-Scale Civil Engineering Research. Test structures can also be subjected to forces with hydraulic equipment to simulate loads experienced in real structures [5].

A Complementary Metal Oxide Semiconductor (CMOS) digital camera is used for close range photogrammetry to measure displacement and strain during tests [6]. The camera-based method was calibrated to determine its accuracy and reliability for a monochrome 8-bit camera (640 by 480) pixels with a 75 mm lens, a view distance of 0.4 m (15 in.), and a field of view of 25 mm by 7 mm (1 in. by 0.75 in.). Displacements were obtained with an accuracy of 5e-3 mm (2e-4 in.). Strain measurements were derived as average strains using three target points and relative displacements between the points. A strain gage was also applied to the back of the test samples at the target points. Overall, the stress-strain curve was captured quite well, including the yield point and plastic strains up to 0.04. The accuracy of the strain measurements was a function of the displacement measurement accuracy and the spacing of the points.

2.3. National Research Council

The National Research Council (NRC) in Canada has a Fire Resistance and Performance of Structures program with fire test facilities that include column, wall, and floor test furnaces, and a large burn hall/smoke tower complex that is 55 m long, 30 m wide, and 12.5 m high. Fire test facilities include a column test furnace with hydraulic jacks that can load along three principal axes and a three-story wall test in the burn hall [7]. NRC is exploring the use of camera-based and fiber-optic systems for measuring structural displacements and strains under fire conditions [8].

2.4. University of Coimbra

A large experimental program on the fire resistance of steel and composite steel and concrete columns with restrained thermal elongation [9] was carried out at the Laboratory of Testing Materials and Structures of the Department of Civil Engineering of the Faculty of Sciences and Technology of the University of Coimbra in Portugal. Their fire test facility has a furnace with a reaction frame for testing structural components and assemblies.

Displacements are measured outside the furnace. A refractory element for high-temperature conditions is welded to the test assembly, and a commercial transducer at the furnace window measures the deformation. Structural strains are not measured inside the furnace. The use of high-temperature strain gages was abandoned after poor outcomes during the Cardington tests (Section 2.6).

2.5. Centre Technique Industriel de la Construction Metallique

The Centre Technique Industriel de la Construction Metallique (CTICM) conducts research to support the steel construction industry in France. The Fire Research Section conducts research projects to improve fire safety and develop computational tools based on fire test programs. CTICM has seven furnaces for fire testing, but also conducts full scale fire tests as needed. CTICM measured temperatures in the heated zone and displacements outside the heated zone during fire tests in a full-scale parking structure; strains were not measured [10].

2.6. British Research Establishment, Cardington Facility

In 1995 and 1996, seven fire tests were carried out on an eight-story, steel-framed building with composite metal deck floors at the Building Research Establishment (BRE) at the Cardington Large Building Test Facility. The displacement and strain response of the structural components subjected to fire conditions were measured primarily outside of the heated zone [11].

Vertical displacements were measured at the top (unexposed) surface of the concrete slab to determine primary beam deflections. Column displacements were measured relative to other (unheated) columns in the building. Rotations at each of the main connections within the test compartments were also measured. Strains were measured with a mix of ambient temperature gages outside of the heated zone and high-temperature gages within the heated zone.

2.7. Architecture and Building Research Institute

The Fire Research Facility of the Architecture and Building Research Institute (ABRI), Taiwan has a large-scale furnace, 4 m × 8 m in cross-section and 5 m in height, to test beam-column assemblies and floors [12]. The furnace is equipped with a 500-ton hydraulic system to apply axial loads to the specimen while it is exposed to fire. Temperatures, axial loads, and axial deformations outside the heated zone are monitored during fire exposure [13].

2.8. Korea Institute of Construction Technology

The Fire Resistance Laboratory of the Korea Institute of Construction Technology (KICT) has several furnaces for building and tunnel fire research [14]. The column furnace can test a column up to 5 m in height and apply a 1000-ton axial load. The furnace for floor systems (slab and beams) can test specimens up to 10 m long and apply loads up to 100 tons. The wall furnace can test wall specimens that are 4 m by 4 m and has a 100 ton load capacity. Temperatures and deformations outside the heated zone are measured during fire exposure [15].

2.9. Summary of Current Practice for Structural Fire Measurement Methods

In structural-fire test facilities around the world, quantitative structural measurements are made outside the heated area, with the exception of a few strain measurements at pre-selected points. The Cardington full-scale fire experiments used many room-temperature and high-temperature strain gages with varying degrees of performance. Other researchers reported that high-temperature gages began to fail at temperatures of 300 °C—temperatures that are well below temperatures reached in fire conditions.

Structures exposed to fire may also have thermally induced loads and failure mechanisms in addition to those that occur at room temperature. However, there is little experimental data to validate analytical models of structural behavior in fire conditions, as there are no reliable methods for measuring displacements and strains across components and subsystems. To advance understanding of how structural systems perform under fire conditions, methods to obtain quantitative measurements of temperature-dependent structural system response need to be developed.

Table 1. Displacement and strain measurement methods in structural fire test facilities.

Test Facility	Displacement Measurements	Strain Measurements
Michigan State University Furnace with applied structural loads	LVDT devices outside of the heated zone Laser-based devices did not work in a fire environment	Not measured, have tried high-temperature strain gages and concrete strain gages
Purdue University Heating elements applied to structural components	CMOS digital camera for 0.4 m (15 in.) focal length	Computed from displacements
NRC, Canada Multiple furnaces; fire laboratory (55 m x 30 m x 12.5 m)	Measured outside heated zone	High-temperature strain gages, fiber-optic sensors
University of Coimbra, Portugal Furnace with reaction frame	Refractory element and camera system	Not measured, have tried high-temperature strain gages
CTICM, France Seven furnaces; parking structures with real fire	Measured outside heated zone	Not measured
BRE Cardington, England Eight-story steel structure with real fire in compartments	Measured outside heated zone	High-temperature strain gages
ABRI, Taiwan Beam/column furnace (7 m x 5 m x 9 m) with a 500 ton actuator	Measured outside heated zone	None listed
KICT Fire Research Center, Korea column, floor, and wall furnaces with axial load capabilities	Measured outside heated zone	None listed

3. Structural Measurement Needs for Fire Conditions

A list of desired performance criteria was developed to guide the review of existing technology and the search for possible innovations that meet the demands of a fire environment. The rationale for the criteria is provided here, and summarized in Table 2. These criteria may not be fully achievable, but even partial fulfillment of the criteria listed in Table 2 will significantly advance structural response measurements in fire.

Table 2. Desired performance criteria for temperature, displacement, and strain sensors.

• Gas temperature measurements from 20 °C up to 1400 °C for a heating duration of four hours with a known resolution and uncertainty
• Structural response measurements in steel and concrete components up to component temperatures of 750 °C
• Method is insensitive to rate of heating and cooling and has the ability to measure in fire environment with smoke and or soot
• Continuous measurement over a surface area or along a line preferred over point measurements
• Displacement measurements that range from 1 mm to 1 m with a resolution of 0.02 mm (0.001 in) to 10 mm (0.4 in)
• Elastic strain measurements that range from 10 μm/m to 2 000 μm/m (0.2 % strain) with a resolution of 1 μm/m to 200 μm/
• Plastic strain measurements from 2000 μm/m (0.2 % strain) up to 200 000 μm/m (20 % strain) with a resolution of 20 μm/m to 2 000 μm/m

Gas temperatures from ambient conditions to 1400 °C should be measured with a known resolution and uncertainty. An upper bound of 1400 °C for gas temperatures is based on expected peak temperatures while burning liquid fuels. Gas temperatures for combustible cellulosic and plastic materials produce peak temperatures in the range of 900 °C to 1100 °C in post-flashover enclosure fires. Pool fires with liquid fuels typically produce 1200 °C temperatures, with an upper bound of 1400 °C. Flame-impingement temperatures may reach 1800 °C, but since the duration is on the order of seconds, there is little effect on the temperature of structural components.

The response of structural components should be measured up to 750 °C for steel and concrete materials. Above 700 °C, structural steel undergoes a eutectoid phase that affects stiffness and yield strength properties. Less than 20 percent of room temperature strength remains above 700 °C [16]. Concrete dehydrates as its temperature rises and retains approximately 20 to 40 percent of its original strength at 750 °C [16].

Measurements of structural response to fire should be insensitive to the rate of heating and cooling, so that the measurement accuracy is only a function of temperature. Fire environments vary during experiments, depending on the fuel type, ventilation, and whether smoke and hot gases are retained in compartments or other enclosed areas. A clean source of heat, such as natural gas, produces a fire without soot or smoke particles. Other fuels, such as heptane, or incomplete combustion of combustible materials produce smoke and soot particles. Hot soot particles re-radiate within the smoke layer and create a hot upper gas layer in compartments. Re-radiation can modify the gas temperatures within the compartment, and consequently, the heating of structural members. Thus, an enclosure fire with cellulosic and plastic contents differs from a quasi-steady-state natural gas or pool fire in the heating rate, peak temperatures, and heating duration of structural components. Therefore, continuous measurements over a surface area or along a line across a component, such as a beam or slab section, would provide a tremendous advantage over point measurements in understanding structural system response to fire.

The range of measurements that may be needed at elevated temperatures is based on a combination of typical ambient responses and possible damage or failure mechanisms. For instance, a floor section with a 10 m to 15 m floor span may initially deflect on the order of a centimeter under gravity loads and then sag at elevated temperatures on the order of a meter. Therefore, the capability to measure displacements from 1 mm to approximately 1 m with a resolution of 0.02 mm (0.001 in) to 10 mm (0.4 in) is desirable. Similarly for strain measurements, linear elastic and plastic strains up to approximately 0.02 are expected for structural response for ambient conditions. Plastic strains may be greatly increased as structural components heat. While the extent of possible plastic strain is not known, an upper bound is expected to be on the order of 0.20. Therefore, the capability to measure linear strain should range from 10 μm/m to 2 000 μm/m (0.2 % strain) with a resolution of 1 μm/m to 200 μm/m. Plastic strain measurements should potentially range from ~2 000 μm/m (0.2 % strain) up to 200 000 μm/m (20 % strain) with a resolution of 20 μm/m to 2 000 μm/m.

Impediments to measuring temperatures, displacement, and strains in a fire environment include issues related to visibility, attachment, and radiation effects. Interference with visual methods of measurement (e.g., camera, laser, infrared, etc.) occurs when wavelengths are reflected or absorbed by smoke particles, distorted by thermal gradients and air column turbulence, or masked by soot deposition on surface marking or tags. Many methods of sensor attachment to structural components at elevated temperatures do not perform well at elevated temperatures. A key issue is the characterization of shear transfer from the structural component to the sensor, through the attachment mechanism. Epoxy adhesives fail above approximately 300 °C. While ceramic adhesives are reported to have higher operating temperatures (1000 °C), they need to be tested in a fire environment. Sensor attachment by welding to steel components or by embedment in a concrete section may be required. Methods may also be needed to thermally protect instrumentation for a range of fire size, duration, and radiative effects.

Equipment costs and total experimental costs should also be considered when developing and evaluating sensors. For instance, disposable (one-time-use) sensors may be desirable for conditions where heat exposure cannot be controlled or is expected to exceed sensor capacity. Ideally, sensor pricing should be moderate compared to the total experimental costs.

4. Candidate Measurement Technologies and Concepts for Fire Conditions

Candidate techniques for measuring temperatures, displacements, and strains in fire conditions are presented. The measurement techniques are described as point, field, or surface methods. Point methods obtain data at a single location with a dedicated data line. Field methods obtain data at multiple points along a data line. Surface methods obtain measurement data across a surface or area.

4.1. Temperature Measurements

4.1.1. Thermocouples

Thermocouples are a well-established technology with known operating temperature ranges and measurement uncertainty. A drawback to thermocouples is that each thermocouple must be individually wired to recording instrumentation. It is not unusual to require hundreds of thermocouples to reasonably measure gas and structural temperatures during a fire test. This method is not scalable to more than several hundred individual measurement points.

4.1.2. Infrared Sensors

Infrared sensors are available as either pyrometers for point measurements or as thermal imagers for surface measurements. Infrared sensors are often used to identify thermal gas plumes during experiments, but such measurements are considered qualitative rather than quantitative.

Most imaging technologies for defense purposes are directed at seeing through smoke, haze, or darkness to identify threats. In general, these threats are warmer than their surroundings, e.g. soldiers at night, or exhaust from vehicle engines obscured by smoke on the battlefield. The requirements for full-scale structures in fire create a different problem—measuring deflections, strains, and temperatures of objects that may be cooler than the surrounding environment with optical distortions due to convection currents of varying temperature and gas density.

Thermal infrared cameras detect certain gases that emit radiation within their spectral range. Long wave infrared (LWIR) detectors detect wavelengths within the (8 to 14) μm bandwidth, which includes most common combustion products. Soot and dust deposition and water deposition/condensation on the thermal imaging optics and target surfaces will affect temperature measurements, and they are difficult to protect against. Soot deposition on the target surfaces will change the emissivity of the surface and, thus, change the temperature measurement. Emissivity also depends on temperature. Nitrogen curtains have been used with limited success. The gases associated with some fuels may interfere with the thermal infrared view of the surface. This is less likely to happen in the LWIR range, as smoke becomes increasingly transparent as the infrared wavelength increases. Very heavy smoke may disturb surface temperature measurements because the thermal infrared will be reflected by a percentage of smoke particles. Water and dust pose similar problems. Flames will probably disturb measurements if they enter the view field [17].

Measurement requirements, such as temperature accuracy, size of the measurement field, or output data type will identify appropriate thermal imaging instruments. Many thermal imagers have a selection of dynamic temperature ranges; it may be necessary to switch between them as the target temperature increases.

Research is being conducted with thermal imaging methods during fire exposure to measure thermal properties of materials, such as emissivity [17]. Such research may lead to thermal methods for quantitative measures of surface temperatures.

4.1.3. Fiber-optic Sensors

Temperature and strain measurements are discussed in this section to avoid repetition, because the same measurement technology is used for both. Fiber-optic sensors are being developed for many applications, including temperature measurements for aerospace applications and strain measurements of dams and bridges. The Naval Research Laboratory (NRL) and Naval Surface Warfare Centers at Norfolk and Carderock have used fiber-optic sensors to monitor strains for detection of cracking in the aluminum superstructure of naval vessels at ambient temperatures. However, there are no reports of research at NRL for fiber-optic sensors at elevated temperatures. Most of the fiber-optic high-temperature applications started with the National Aeronautics and Space Administration (NASA) in the late 1990s to monitor strains in aircraft and spacecraft for temperatures up to approximately 300 °C. Fiber-optic sensors are also used on bridges and other structures to monitor temperatures (less than 300 °C) or strains during construction or for health monitoring. Examples include monitoring temperatures in large concrete sections while curing, measuring cable and pavement temperatures in bridges, and monitoring strains of pipelines in landslide areas [18].

Optical fibers are silica glass doped with small amounts of germanium and theoretically can function in temperatures up to the glass transition temperature of 1200 °C. A cladding is formed around the optical fiber that protects it from damage and moisture and forms a waveguide around the fiber with a lower refractive index. Coatings are applied over the cladding to provide service life durability and strength. Most coatings in use today are limited to temperatures of approximately 300 °C. There are specialty coatings, such as gold-based coatings, that reportedly can reach 700 °C. However, in a laboratory setting, fiber-optic coatings may not be needed for one-time experimental applications [19].

Fiber-optic sensors that measure changes in temperature or strain are based on three forms of scattered light. A laser light at a specific wavelength is pulsed into an optical fiber, and is scattered as it travels down the fiber by interactions between the photons and the crystalline structure of the glass fiber. This scattering takes three forms: Rayleigh, Brillioun, and Raman scattering. Rayleigh scattering is the result of elastic collisions, and the scattered (reflected) signal has the same wavelength as the incident light. Brillioun and Raman spectra scattering are both inelastic events, and result in scattered signals comprising the red-shifted Stokes and the blue-shifted anti-Stokes components. In Brillioun scattering, the amplitudes of the Stokes and anti-Stokes signals are predictable given the amplitude of the incident light, but the wavelengths of the resulting signals are variable. In contrast, the Raman spectra Stokes and anti-Stokes signals have predictable wavelengths, but the relative amplitude of the two signals varies with temperature [20].

Fiber-optic sensing mechanisms used for structural measurements include Extrinsic Fabry-Perot Interferometer (EFPI), Fiber Bragg Gratings (FBG), Long Period Fiber Grating (LPFG), and Distributed Sensing System (DSS). EFPI sensors have a gap along the fiber length and are a point measurement method. As the gap length changes due to temperature or strain effects, the interference pattern for the reflected light signals changes. The maximum temperature for EFPI sensors is approximately 1000 °C, since the sensor is based on a gap in the fiber [21]. FBG sensors have a series of spaced lines etched into the optical fiber to reflect the light signal at a given frequency. Changes in fiber length, whether due to temperature or strain, cause a shift in the reflected frequency. The maximum temperature for FBG sensors is approximately 600 °C, due to the germanium doping typically used with optical fibers. If the FBG area is heat treated (annealed), the sensor may perform up to 1100 °C [22]. LPFG sensor gratings are fabricated by exposing the fiber core to ultraviolet (UV) radiation. This technology has been used to make Bragg-type fiber gratings (wavelengths less than a micrometer) and LPFG gratings (wavelengths in the hundreds of micrometers) [23]. The gratings couple the guided fundamental mode in a single-mode fiber to forward propagating cladding modes. These modes decay rapidly as they propagate through scattering losses at the cladding-air interface. Since the coupling is wavelength-selective, the fiber grating acts as a wavelength dependent loss element [24]. DSS sensors measure temperature or strain anywhere along the fiber length. The measurement technique is based on low level backscattering of the laser signal at minor variations in the fiber density.

DSS techniques that are based on Raman or Rayleigh scatter measurement typically employ optical frequency domain reflectometry (OFDR) for applications requiring higher spatial resolution. A commercially available fiber-optic system based on Rayleigh OFDR measures temperatures (and strains) at user-selected points along a 70 m fiber, with reported spatial resolution on the order of 1 cm and uncertainties of ± 1 °C [25]. A second commercially available fiber-optic system based on Raman OFDR [26] measures temperatures along a fiber up to 2 km long with possible resolutions of spatial sampling (0.25 m minimum) and measurement time per fiber (5 s minimum), depending on several parameters, such as the length and type of fiber.

DSS techniques based on Raman and Brillouin scatter measurements typically employ optical time domain reflectometry (OTDR) and may not be well suited for applications requiring high spatial resolution. Brillouin techniques are often used for long distances on the order of kilometers to monitor temperatures over a region on the order of meters [27]. However, a new Brillioun measurement technique has high-spatial-resolution and long-range distributed temperature sensor, based on experimental results [28]. It uses a differential pulse-width pair Brillouin optical time-domain analysis (DPP-BOTDA), which detects differential Brillouin gain instead of Brillouin gain itself. As a result, the spatial resolution of the Brillouin technology is comparable to Rayleigh OFDR for a much longer fiber length. The reported resolution of BOTDA technology for the combined sensing length and spatial resolution is 2 cm spatial resolution over a 2 km sensing length [28].

4.1.4. Summary of Temperature Measurement Technology

Table 3 summarizes current technology for temperature measurements. Thermocouples, infrared-based methods, and fiber-optic techniques have been used to measure gas and structural temperatures, and are well established technologies for non-fire environments. The measurement accuracy and uncertainty of each method needs to be determined for a fire environment. Lonnermark, et. al. [29] compared thermocouple and fiber-optic temperature measurements during an experimental tunnel fire with gas temperatures up to

250 °C. Gas temperatures measured by thermocouples with two bead diameters (0.5 mm and 2 mm) and by a FBG sensor were compared with a gas temperature history based on a steady-state heat transfer model for two fire tests in the model-scale tunnel. The temperature measured with the FBG sensor and the calculated gas temperature were within approximately 25 °C, while the temperatures measured by the thermocouples were significantly lower by up to 75 °C. These results demonstrate the need for further research of thermocouple and fiber-optic gas temperature measurements in real fire conditions with time-varying gas temperatures.

Table 3. Technology for temperature measurement.

Measurement Type	Technology	Technical Issues for Fire Conditions
Point	Thermocouples	Established technology with known operating temperature ranges and measurement uncertainty. Measurement uncertainty may vary with temperature.
Point	Infrared pyrometer	Infrared gun aims with laser, measures up to 1600 °C. Generally used for qualitative measurements. Subject to optical interference.
Surface	Thermal Imaging	Thermal imaging with video, 3 frames/s, measures up to 1000 °C. Generally used for qualitative measurements. Subject to optical interference.
Field	Fiber-optic Sensor	Need to characterize measurement resolution and uncertainty for varying temperature and fiber conditions.
Field	Fiber-optic Sensor	Need to characterize measurement resolution and uncertainty for varying temperature and fiber conditions.

4.2. Displacement Measurements

4.2.1. Linear Transducers

Traditional structural displacement measurement techniques include a linear position transducer (LPT, also referred to as draw wire or string pot devices) and a linear variable differential transformer (LVDT). The LPT measures the amount of wire drawn from a spool during structural deformations and displacements. The LVDT measures translational displacements through induced voltage changes as the core moves through electromagnetic coils. Use of LPT and LVDT in fire conditions would require thermal protection or materials suitable for exposure to elevated temperatures (e.g., no significant thermal expansion of draw wire for temperature range of interest).

4.2.2. Lasers

Lasers are widely used for noncontact measurements of distances to objects and changes in position. Laser rangefinder devices, with appropriate software, can define three-dimensional objects such as buildings or terrain. However, lasers perform poorly in fire conditions because the light spectrum for flames distorts or masks the laser light frequencies. Suitable laser frequencies and signal processing techniques for their use in fire environments need to be identified.

4.2.3. Digital Image Correlation (DIC) Systems

Digital image correlation (DIC) systems process sequential images to compute surface measurements of displacement and strain. Digital images can record the displacement of thousands of data points over a component surface. The recorded data history can be reprocessed after the experiment to correct for rigid body motions and out-of-plane curvatures. Measurement over areas rather than at discrete points could reduce the need for additional experiments.

For digital image correlation measurements, a coating with a random pattern is applied to a component surface. The digital camera takes images during the test. Software processes the images for relative movement of the random pattern, and surface strains are computed between two points based on relative displacements. Coatings have been tested that adhere up to 800 °C in a furnace. However, smoke or soot may interfere with imaging of the coated surface.

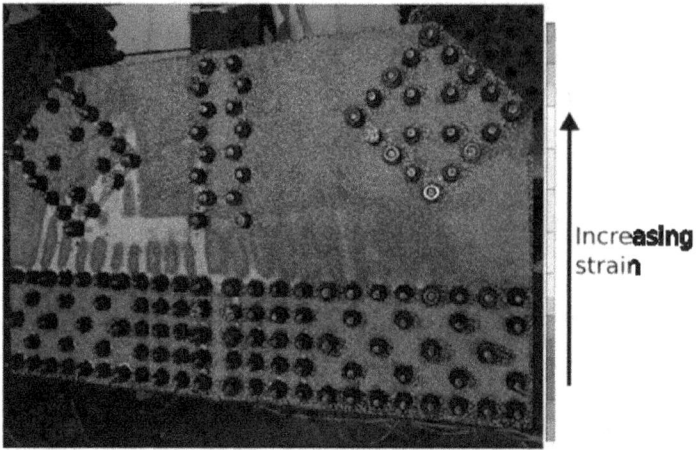

Figure 2 Example of DIC computed strains on a bridge plate
with surface markers and superimposed strains.

NIST researchers have applied this technique for gusset plates at room temperature, as shown in Figure 2, in collaboration with the Federal Highway Administration to measure strains at ambient conditions [30]. The figure demonstrates the state of the art in structural strain measurement for ambient conditions. In the figure, the red color denotes regions of high strain. The heterogeneous distribution of strains would be impossible to capture or understand if only a dozen strain gages had been applied to the plate. Because image-based techniques do not choose discrete locations for measurements, it is possible to analyze any area in the field of view after the test.

Digital Image Correlation (DIC) is an image analysis technique, that is not limited to optical wavelengths. Any set of images, not just those formed by visible light, is amenable to the technique. However, infrared wavelengths are probably not feasible, since as the temperature increases, the intensity of the background radiation increases. Some success has recently been reported with short wavelength illumination ($\lambda = 450$ nm) and band-pass filtering all wavelengths outside the range ($425 < \lambda < 470$) nm [31]. This approach attempted to measure the coefficient of thermal expansion of stainless steel in a resistance-heated (i.e., no flames) furnace. A significant challenge in extending the technique to the fire facility will be creating sufficient short wavelength illumination on the target. Most previous research on high-temperature digital image correlation has focused on imaging small parts in furnaces, rather than a building-sized element.

Use of DIC techniques in fire conditions will require identifying methods to increase the contrast of the image under fire conditions. The contrast of surface coating patterns can be increased by developing markings that fluoresce in ultraviolet wavelengths and by filtering the reflected light to include only the wavelengths from the fluorescent markers. The technical challenge lies in developing coatings or paints that both fluoresce and adhere to the specimen in fire conditions.

Image processing methods developed for inspections of underwater structures, such as pipelines and offshore platforms where visibility is poor due to turbidity and debris, may be applicable to post-processing of images with smoke and visual disturbances in a fire environment [32]. Innovations based on other digital imaging technologies could also be employed, such as adaptive optics. Adaptive optics have been applied to a number of imaging applications to remove the effect of disturbances in the visual medium, such as astronomical telescopes, greenhouse gas detection, and free-space communication.

4.2.4. Summary of Displacement Measurement Technology

Table 4 lists technologies for displacement measurements and technical issues requiring research for use in fire conditions. Noncontact technologies, such as digital imaging or laser-based systems, provide many measurements without wiring from each measurement location. Noncontact technology can potentially save considerable labor and time in the construction and setup of an experiment, and provide a richer data base for validation of analysis models. Depending on the number of components and complexity of the structural system, more than one of these methods may be required in a given test.

The following challenges must be addressed to use digital image correlation (DIC) techniques in fire conditions:

(1) Develop coating systems that increase image contrast with coatings or paints that both fluoresce and adhere to the specimen.

(2) Determine optimal imaging measurement methods (e.g., short wavelength illumination on large targets, filtered wavelengths, laser, x-ray) and image quality for multiple angles, complex shapes, and fields of view.

(3) Optimize digital image processing for measurement of local deflections and strains that resolve interference issues that include air turbulence, smoke and soot obstruction of target markings, and

fire spectrum interference with image lighting and resolution.

(4)   Quantify measurement uncertainty as a function of temperature.

Table 4. Measurement issues for displacement measurements.

Measurement Type	Technology	Measurement Issues for Fire Conditions
Point	LPT LVDT	Thermal protection or material modifications to sensor is needed for exposure to gas temperatures up to 1400 °C.
Point	Lasers	Laser frequencies and signal processing techniques need to be developed for use in fire environments.
Surface	DIC	Methods to increase image contrast, short wavelength illumination on large targets, and image processing for smoke and turbulence in fire conditions are needed. Also need to characterize measurement resolution and uncertainty for varying fire conditions.

4.3.   Strain Measurements

4.3.1.   Strain Gages

Strain gages measure linear strains at discrete points. Measurement science for strain gages is well understood, but still has significant performance issues in high-temperature applications. Research is being conducted to identify materials that are thermally stable at elevated temperatures [33]. Examples of high-temperature strain gages for steel components include an iron-chromium-aluminum alloy wire resistance strain gage of the Hoskins 875 alloy that has a temperature range up to 850 °C and a 0.01 strain limit. Users are cautioned that bonding the strain gage to a material with a higher thermal expansion rate will induce tensile stresses and to avoid rapid heating or cooling (rates greater than 14 °C/s) that will produce transient apparent strains [34]. This type of strain gage was used in a digital image correlation demonstration test; see Section 5.

Epoxy bond materials that attach strain gages to steel surfaces can be used up to temperatures of approximately 400 °C. At higher temperatures, either ceramic bond materials or welds must be used. Strain gages for concrete sections can be bonded to the concrete surface (for crack detection), bonded to the reinforcement, or embedded in the concrete section for strain measurements up to 100 °C.

As noted in Section 5, the performance of high-temperature strain gages in furnace and fire conditions is highly variable and unreliable. Further research is needed to identify the sources of poor performance and develop improved strain gages for fire conditions.

4.3.2. Digital Image Correlation

Digital image correlation (DIC) systems (see Section 4.2.3) compute surface strains using measurements of relative displacement between points. All the limitations of displacement measurement apply to strain measurement.

4.3.3. Fiber-optic Sensors

Fiber-optic sensors measure strains along a fiber length using FBG, EFPI, or LPFG sensors or DSS sensing (see Section 4.1.3). The optical fiber must be either attached to the surface of a steel component or embedded in a concrete section. Similar to strain gages, attachment methods other than epoxy bonds are required at elevated temperatures. Attachment methods between the fiber-optic sensor and/or fiber and the steel or concrete component are not well developed for fire conditions. Instead, each application requires development of an attachment method that considers issues such as material properties and differential thermal expansion at elevated temperatures [19]. The attachment method must be developed so that the transfer of strain from the steel or concrete component through the bonding agent and cladding to the core can be characterized.

Simultaneous strain and temperature measurements were investigated for a LPFG sensor with an experimental strain transfer mechanism designed for large strains [19]. The strain transfer mechanism included both gauge length change and shear lag effects, and measured strains up to 15 200 μm/m in temperatures up to 700 °C. At higher temperatures, different adhesives must be used to bond the optical fiber, the protective sleeve, and the substrate.

4.3.4. Summary of Strain Measurement Technology

Table 5 lists technologies for strain measurements and technical issues requiring research for use in fire conditions. Strain gages can provide point measurements to confirm surface or distributed strain measurements if their performance issues in fire conditions are resolved. Digital image correlation challenges are the same as those given in Section 4.2.3.

Three challenges must be addressed to use fiber-optic sensor techniques in fire conditions:

(1) Determine the best fiber-optic sensing method for temperature and strain measurements across large (beam length) and small (connection) scales in fire conditions. Possible fiber-optic sensors include Fiber Bragg Gratings (FBG), Extrinsic Fabry-Perot Interferometer (EFPI), Long Period Fiber Grating (LPFG), and Distributed Sensing System (DSS).

(2) Develop methodologies for bonding, embedment, or encasements of fiber-optic sensors that measure large strain measurements in fire conditions. Analytically determine and experimentally validate the strain transfer from the component to the sensor.

(3) Quantify measurement uncertainty as a function of temperature.

Table 5. Measurement issues for strain measurements.

Measurement Type	Technology	Measurement Issues for Fire Conditions
Point	Strain Gage	Further research is needed to identify the sources of poor performance and develop improved strain gages for fire conditions.
Point	Concrete Strain Gage	Thermal protection for concrete strain gages or new measurement techniques at elevated temperatures need to be developed. Strain transfer from the component to the sensor need to be characterized.
Surface	DIC	Methods to increase image contrast, short wavelength illumination on large targets, and image processing for smoke and turbulence in fire conditions are needed. Also need to characterize measurement resolution and uncertainty in for varying fire conditions.
Points along a fiber; Continuous along fiber	Fiber-optic Sensor	Sensor integrity and attachment methods in fire conditions need to be developed. Strain transfer from the component to the sensor need to be characterized.

5. Digital Image Correlation and Strain Gage Demonstration Test

5.1. Background and Goals

To evaluate the suitability of digital image correlation (DIC) for strain and displacement measurement, a demonstration test in the NIST Large Fire Facility was conducted. The demonstration test had three goals:

1. Evaluate the suitability of digital image correlation for strain measurement under fire conditions.

2. Compare digital image correlation to traditional high-temperature strain-gage measurements.

3. Understand the sensitivity and uncertainty of both digital image correlation and high-temperature resistance strain gages in real fire conditions, as opposed to furnace conditions.

Appendix 1 details the configuration and results of the test. This section summarizes the important results and major findings; see Appendix 1 for more details.

5.2. Summary of the Demonstration Test

In preparation for the test, a suitable paint and method for applying the paint to a test specimen of mild steel plate was identified. Three candidate paints and preparation techniques were first evaluated in a one-hour furnace exposure test. Two survived exposures of 600 °C, with different amounts of discoloration. None survived for an hour at 700 °C. The most durable and color-fast coating was then tested under direct flame exposure at temperatures up to 700 °C for short times to demonstrate that the paint adhered.

A three-phase test with the prepared test plate was conducted in the Large Fire Facility. Rigid body displacements were used to determine the DIC measurement uncertainty at ambient conditions, and thermal strains were measured with DIC methods and strain gages. No mechanical loads were applied to the test specimen during the tests.

5.2.1. Test Configuration

The test specimen consisted of a 30 cm x 45 cm x 1 cm mild steel plate instrumented with two high-temperature strain gages and eight Type K thermocouples. The side opposite the strain gages was patterned with black overspray on a white background for strain measurement by digital image correlation. The plate was suspended about 1 m above a natural gas burner capable of 70 kW output. Figure 3 shows the instrumented plate with its black and white patterning, suspended above the gas burner. The vantage point is just behind one of the two cameras. The thermocouples and strain gages are on the opposite side of the plate.

Figure 3. Configuration during the demonstration test seen from just behind the cameras, looking at the random paint pattern applied to the test specimen.

5.2.2. Results

Phase 1 of the test produced an estimate of the absolute uncertainty in the strain. In phase 1, the plate was translated left and right (x-direction) as well as up and down (y-direction) with no fire present. Since rigid-body motion should produce zero strain, this test established the uncertainty in strain measurements with DIC at ±50 μm/m. This strain corresponds to the strain produced by a room-temperature elastic loading of 10 MPa or a thermal expansion strain produced by a 4 °C temperature change.

In Phase 2 of the test, the burner was placed between the plate and the cameras, to assess the effect of air turbulence on the DIC strain uncertainty. Again, the plate was translated in the x- and y- directions. The reading-to-reading scatter in strains was 150 times larger than in Phase 1, but the mean strain calculated was still zero.

In Phase 3, the plate was located directly over the flame. A major limitation of the test was that the 70 kW fire was only able to heat the plate to about 200 °C in the open configuration (no compartment to collect heated gases), which is well below the expected temperatures in structures during fire conditions. Figure 4 shows DIC computed thermal strains in the horizontal (x-x) and vertical (y-y) directions of the plate as a function of temperature during phase 3 of the test. The dashed line is a linearization of the accepted value [35] for thermal expansion based on temperature. The dash-dot line is the linear regression of the data. Over the range of the data, the slopes of the lines differ by less than 1.9 %, and the maximum difference between the literature and fitted strain values is less than 11 %.

The high-temperature strain gages are designed to self-compensate for thermal strains; see Appendix 1 for details. Unfortunately, both gages displayed large *negative* apparent strains during all phases of the test. The magnitude of the apparent strain-gage strain was about 25 percent of the DIC-measured thermal strain. Subsequent testing of the strain gages in a furnace demonstrated that these apparent strains were quite repeatable and proportional to temperature. Extensive discussions with the strain gage manufacturer did not lead to any method for compensating for these strains. The demonstration test confirmed the reported difficulties with obtaining reliable measurements with high-temperature strain gages.

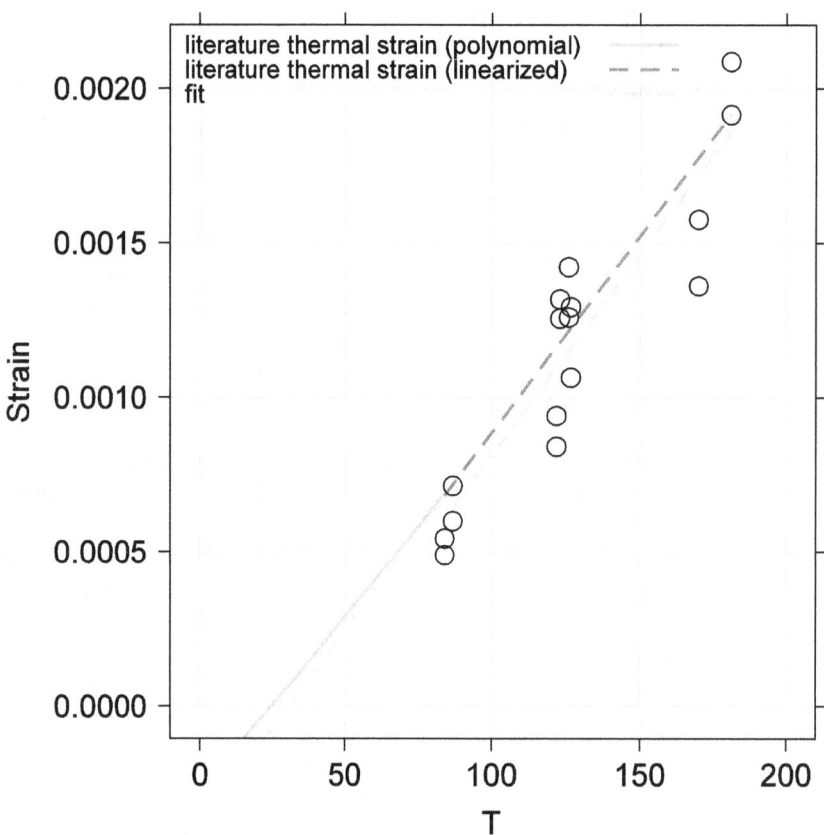

Figure 4. Comparison of the thermal strains, *e*, measured by DIC as a function of temperature near the two strain gages during Phase 3 of the test. Dashed line is the computed value for thermal expansion based on temperature. Dash/dot line is the linear regression of the data

5.3.   Summary of Demonstration Test Findings, Conclusions, and Recommendations

Although the test plate did not reach the temperatures that steel components can reach in fire conditions, the results of the demonstration test encouragingly support the feasibility of digital image correlation methods for measuring strain in fire conditions. The demonstration test produced four significant findings.

1. Commercial high-temperature paint can survive up to 600 °C for an approximately 1 h furnace exposure.

2. Rigid body translations produced no apparent strain for a plate viewed through air turbulence over a fire.

3. Digital image correlation measured the thermal strain of a plate heated by fire to within 11 % of the accepted value.

4. Conventional high-temperature strain gages produced apparent compressive strains on heating.

Based on these findings, digital image correlation methods show significant promise for measuring strains and deformations in fire, and could revolutionize structural fire measurements. The data acquisition and analysis of the images employed no special illumination techniques or image post-processing. Both of these approaches have significant potential to improve the quality of the measurements. Despite decades of development, high-temperature strain gages are still difficult to use, and have poorly understood thermal response.

Recommendations for developing digital image correlation as a strain-measurement tool for the National Fire Research Laboratory include the following steps:

1. Conduct tests that heat the plate to more than 600 °C to further study the adherence of commercial paints in fire conditions.

2. Evaluate methods for post-processing the digital image correlation data to reduce the uncertainty in strains.

3. Add mechanical loads and/or deformations to the fire tests.

4. Develop quantifiable uncertainties for strain-measurement methods.

6.   Candidate Technologies for Next-generation Measurements of Structural Response in Fire

Based on the review of available technology and their measurement challenges, and the demonstration test with digital image correlation and high-temperature strain gages, digital image correlation and fiber-optic technologies appear to be the most promising technologies for temperature, displacement, and strain measurements of structural systems in fire conditions. Fiber-optic and digital image correlation measurement capabilities are complementary technologies. Digital image correlation methods measure surface displacements and strains. Fiber-optic sensors measure strain along a fiber on any exterior surface or in the interior of concrete sections. Parallel fiber-optic sensors measure corresponding temperatures for strain measurements.

Advances in computer processing have led to distributed systems that employ image techniques or fiber-optic sensors to measure strains and deformations at hundreds to thousands of points on a structure. Although these techniques have been successfully applied at room temperature, significant challenges remain to apply them to the extremely hostile environment of a fire (e.g., flames, smoke, soot, high temperatures, radiation, and air turbulence).

Image-based techniques, using digital image correlation, measure the strain or displacement between successive images of a suitably patterned structure. Because image-based techniques do not choose discrete locations for measurements, it is possible to analyze any area in the field of view after the test. Because fire tests are expensive and time consuming, this post-test analysis capability is particularly significant. Table 6 lists the challenges and possible solutions for applying digital image correlation methods for measurements in fire conditions in increasing order of difficulty. Issues with smoke and soot are likely to be the greatest challenge.

Today, fiber-optic sensors are used to monitor the structural health of buildings and infrastructure. NASA started work in the 1990s to use fiber-optic sensors at elevated temperatures (up to 300 °C) to monitor strains in aircraft and spacecraft. In contrast, very little work has been done at temperatures found in fires because the measurement science challenges are significant. For example, one commonly used fiber-optic sensor is based on a Fiber Bragg Grating (FBG). In this type of sensor, the grating reflects a particular wavelength of light. However, the wavelength can shift due to changes in both temperature and stress on the grating. In a fire scenario to measure structural performance, these two effects cannot be separated.

However, using a Raman scattering technique to measure the temperature of the fiber, it may be possible to extract the strain-induced component from the FBG signal. Other fiber-optic sensors that could be investigated for strain measurements include Extrinsic Fabry-Perot Interferometers (EFPI), Long Period Fiber Gratings (LPFG), and Distributed Sensing Systems. In addition to potential signal confusion at elevated temperatures, fiber-optic sensors must overcome the effects of differential thermal expansion between materials, large strains that exceed the fiber strain limit, and possible measurement artifacts from various attachment methods due to high-temperature effects. Table 7 lists the challenges and possible solutions for using fiber-optic sensors for measurements in fire conditions.

Table 6. Challenges and possible solutions for digital image correlation.

Challenge	Possible solutions
Integrity and adherence of the pattern at high temperature and under fire assault	High-temperature paints Anodized coatings Flame- or plasma-sprayed ceramic coatings
Optical radiation from fire overwhelms image contrast; need to increase image contrast	Infrared filtering Short wavelength/UV illumination Monochromatic, high-intensity illumination
Turbulent variable-density air currents change the optical path and distort the image, leading to spurious strain measurements	Integration of IR and visible spectrum images Electro-magnetic wave propagation using Mie scattering and Kolmogorov turbulence theory Data-driven modeling from integrated images to correct distorted measurements Adaptive optics
Smoke, flame, and soot obscure pattern	Image post-processing techniques, for example Retinex

Point measurement technologies should also be evaluated and developed to provide a secondary level of measurements. However, there are significant disadvantages to relying solely on point measurement methods. Discrete measurement devices require separate data lines for each sensor, and there can be significant associated labor and costs. More significantly, point measurements may miss critical displacement or strain locations, reducing the usefulness of expensive laboratory experiments, and potentially resulting in misleading conclusions.

In summary, traditional room-temperature structural experiments measure displacements and strains at a few "critical" nodes, selected *a priori*. In fire conditions, existing measurement methods fail due to either high-temperature effects on materials or optical saturation from the fire. Notably, high-temperature strain gages have shown a lack of reliability, based on reports from other laboratories and the experiment at NIST. Rather than improve point measurement methods, research should focus on developing innovative distributed measurement techniques for fire environments based on fiber-optic sensors and digital image correlation. The ability to measure deformations and strains inside, rather than outside, the fire zone would greatly enhance structural performance measurements in the newly commissioned NFRL. Experimental data are critical for designing safer buildings, bridges, and other infrastructure using validated next generation computational tools.

Table 7. Challenges and possible solutions for fiber-optic measurements.

Challenge	Possible solutions
Ensuring fiber and sensor integrity and performance at large strains using materials with different thermal expansion behavior, over a large range of temperatures	Characterize performance of existing sensors at elevated temperatures (e.g., FBG, EFPI , LPFG, DSS). Novel, modified, or hybrid fiber sensor types to measure temperature and strain in fire conditions that optimize fiber and sensor characteristics, such as an external Fabry-Perot cavity
Shielding fiber-optic sensors from fire radiation and gas temperatures	Insulated covers with a small foot print to minimize interference with the heating of components (mineral wool, ceramic, etc.)
Deconvolving strain and temperature measurements	Accurate measurements of temperature along fiber to separate thermal and mechanical strains
Developing attachment methods	Ceramic adhesives Welded or mechanical attachment

7.   Proposed Development Plan

The following research plan is outlined for the development and/or adaptation of existing technologies for measurements in fire conditions with fiber-optic and digital imaging systems. The successful measurement methodology must address the effects of high temperatures on material properties and optical distortion or obscuration by thermal plumes, soot, and smoke.

Four critical challenges for digital image correlation (DIC) techniques in fire conditions have been identified.

1. Developing coating systems that increase image contrast with coatings or paints that both fluoresce and adhere to the specimen.

2. Determining optimal imaging measurement methods (e.g., short wavelength illumination on large targets, filtered wavelengths, laser, x-ray) and image quality for multiple angles, complex shapes, and fields of view.

3. Optimizing digital image processing for measurement of local deflections and strains that resolve interference issues that include air turbulence, smoke and soot obstruction of target markings, and fire-spectrum interference with image lighting and resolution.

4. Quantifying measurement uncertainty as a function of temperature and fire conditions (e.g., fuel types).

Three critical challenges for fiber-optic sensor techniques in fire conditions have been identified:

1. Determining the optimal fiber-optic sensing method for temperature and strain measurements across large (beam length) and small (connection) scales in fire conditions. Possible fiber-optic sensing technology includes Fiber Bragg Gratings (FBG), Extrinsic Fabry-Perot Interferometer (EFPI), Long Period Fiber Grating (LPFG), and Distributed Sensing System (DSS).

2. Developing methodologies for bonding, embedment, or encasements of fiber-optic sensors for large strain measurements in fire conditions. Analytically determine and experimentally validate strain transfer from the structural component to the sensor.

3. Quantifying measurement uncertainty as a function of temperature.

The proposed research starts with tests of existing sensors and technology to determine their limitations, bounds of performance, and measurement uncertainty for ambient and fire conditions. The research will be iterative in nature, with progressive development of successful methodologies, technologies, and materials suitable for fire conditions.

Stage 1: *Develop measurement science in quiescent, high-temperature conditions for simple deformations and identify uncertainty sources.* Investigate the technical challenges listed in Table 6 and Table 7 and develop solutions for accurate, reliable sensor measurements in a fire environment. Use small-scale, clean-burning fires or furnaces in the early stages of research. For digital image correlation, develop methods to integrate infrared and visible spectrum images to model distortion of the optical path to ensure metrological quality of the visible images for rigorous digital image correlation.

Research goal: Develop the supporting science and technology for measuring deformations of a small, simple test structure with quantified uncertainty under elevated temperature conditions.

Stage 2: *Extend measurement technology to structural systems in clean fires.* Extend the measurement technology to a model structure, such as a steel or concrete frame with realistic loads, under clean fire conditions (no significant smoke or soot). Study the quality of thermal and load measurements with digital image correlation and fiber sensors, as well as air column distortions for digital image correlation.

Research goal: Extend the supporting science and technology for measuring deformations of a realistic test structure with quantified uncertainty under clean fire conditions.

Stage 3: *Extend measurement technology to structural systems in dirty fires.* Continue developing measurement technology to experiments under dirty fire conditions (shown in Figure 5), adding varying levels of smoke or soot to digital image correlation tests. Continue developing fiber-optic sensors with a focus on robust attachment approaches. Apply electro-magnetic wave propagation models using Mie scattering and Kolmogorov turbulence theory to reduce the influence of smoke on soot in the image quality.

Research goal: Develop the supporting science and technology for measuring deformations of a realistic test structure with quantified uncertainty under dirty fire conditions.

Stage 4: Evaluate refined fiber-optic and digital image correlation measurement techniques and methodology to quantify sources of uncertainty.

Evaluate measurements of several steel and concrete beams that have mechanical loading, thermal restraint at the end supports, and staggered placement that creates difficult surfaces for viewing, and are located inside a compartment to allow for assessment of compartment fire conditions (e.g., hot zone, smoke, and soot).

Research goal: Measure the deformation of a real structural system with complicated load paths, restraints, and internal areas with quantified uncertainty in realistic fire conditions.

Figure 5. NIST compartment fire experiment with a sooty 4 MW fire.

34

8. Summary

At present, temperatures, displacements, and strains of structures in fire conditions are measured with single point sensors, such as thermocouples, transducers outside the heated zone, and high-temperature strain gages. Significant advancement of measurement science for structures in fire conditions are needed to support the validation of analytical tools and performance-based-design methodologies.

A set of desired performance criteria for structural-fire measurements were established to guide the assessment and development of candidate technologies and measurement systems:

- Measurement methods that capture the response of an entire surface.

- Sensors that function at temperatures as high as 750 °C, and are insensitive to heating rate.

- Displacement measurements ranging from 1 mm to 1 m.

- Elastic strain measurements up to $e = 0.002$ with a resolution of ± 200 μm/m and plastic strain measurements up to $e = 0.2$, with a resolution better than ± 2000 μm/m.

Two innovative measurement techniques, distributed fiber-optic sensors and digital image correlation field measurements, will enable full characterization of structural elements, assemblies, and systems experiencing mechanical and transient thermal loads if the significant technical challenges identified in this report can be resolved. A high density of high-quality test data is essential for validating the next generation of design tools for the practicing engineer.

A four-stage development plan is proposed to overcome these challenges. The early stage develops the measurement science with simple heating and deformations. The intermediate stage employs more realistic fires and more complicated deformations. In the final stage, the measurement uncertainty of both methods will be quantified.

The ability to measure strain and displacement of structural components and systems during fire experiments is critical for (1) obtaining experimental data to validate structural analytical models and (2) developing performance-based design methods for structures in fire. The successful completion of the proposed work will:

- Advance the state-of-the-art in structural fire measurements from a few unreliable point measurements to full-field measurements with documented uncertainty.

- Provide a new measurement capability for other industries, such as aircraft, spacecraft, ships, and nuclear facilities, with fire hazards or elevated temperatures that does not presently exist.

- Lead to advances in safer buildings, bridges, and other infrastructure with quantified structural fire resistance and safety factors for fire.

- Lead to new experimental methods, validated design and computational tools, and best practices for structural fire laboratory measurements.

- Advance the capabilities of the NFRL.

9.   References

1. Beitel J, Iwankiw N. Analysis of Needs and Existing Capabilities for Full-Scale Fire Resistance Testing. Gaithersburg, Md: National Institute of Standards and Technology; 2008. Report No.: GCR 02-843-1 http://fire.nist.gov/bfrlpubs/fire02/art028.html.

2. The Center for Structural Fire Engineering and Diagnostics (SAFE-D). [Online].; 2012 [cited 2012 7 24. Available from: http://www.egr.msu.edu/cee/research/fire_center.pdf.

3. Dwaikat MMS, Kodur VKR, Quiel SE, Garlock ME. Experimental behavior of steel beam-columns subjected to fire-induced thermal gradients. *Journal of Constructional Steel Research,* 67:30-38, 2011.http://dx.doi.org/10.1016/j.jcsr.2010.07.007.

4. Kodur VKR. Guidelines for Improving the Standard Fire Resistance Test Specification. *Journal of ASTM International,* 6(7), 2009.http://dx.doi.org/10.1520/JAI102275.

5. Bowen Laboratory for Large-Scale Civil Engineering Research. [Online]. [cited 2012 7 24. Available from: http://www.purdue.edu/newsroom/research/2011/110901Varma911.html.

6. Selden K. Evaluation of a CMOS digital camera for structural engineering measurement. Research Report. West Lafayette, IN: Purdue University, Bowen Laboratory; 2010. Report No.: 2010-03.

7. National Research Council Canada, Institute for Research in Construction. [Online]. [cited 2012 07 24. Available from: http://www.nrc-cnrc.gc.ca/eng/solutions/facilities/fire_resistance.html.

8. Adelzadeh M, Green, MF, Khalifa T, Li W, Bao X, Beni N. Fibre optic sensors for high-temperatures and fire scenarios. National Research Council, Institute for Research in Construction, Canada.; 2011. Report No.: NRCC-53921 http://www.nrc-cnrc.gc.ca/obj/irc/doc/pubs/nrcc53921.pdf.

9. Carvalho LT, Rodrigues JPC, Landesmann A, Laim L. Numerical simulation of steel and composite steel and concrete columns in fire. In Nordic Steel Construction Conference; 2009. http://www.nordicsteel2009.se/pdf/59.pdf.

10. Joyeux D, Kruppa J, Cajot LG, Schleich JB, van de Leur P, Twilt L. Demonstration of real fire tests in car parks and high buildings. Brussels, Belgium:; 2002. Report No.: EUR 20466 EN http://bookshop.europa.eu/en/demonstration-of-real-fire-tests-in-car-

parks-and-high-buildings-pbKINA20466/.

11. Kirby BR. The Behaviour of a Multi-Storey Steel Framed Building Subject to Fire Attack: Experimental Data. British Steel Swindon Technology Centre, United Kingdom; 1998. http://www.mace.manchester.ac.uk/project/research/structures/strucfire/DataBase/TestData/BRETest/BehaviourMultiStoreySteelBuilding.pdf.

12. Chen YH, Chang YF, Yao GC, Sheu MS. Fire Damaged RC Columns Subjected to Biaxial Bending. In 32nd Conference on Our World in Concrete & Structures, 28 - 29 August 2007, Singapore; 2007. http://www.cipremier.com/e107_files/downloads/Papers/100/32/100032021.pdf.

13. Fang IK, Sullivan PJE, Lee Cc, Fang IC, Yeh TY, Wu MY. Fire resistance of beam-column subassemblage. *ACI Structural Journal,* 109(1):31-40, 2012.

14. Korea Institute of Construction Technology. [Online].; 2012 [cited 2012 7 24. Available from: http://www.kict.re.kr/eng/fac/fire.asp.

15. Park S, Kim H, Kim H, Hong K. Fire Resistance of the Korean Asymmetric Slim Floor Beam Depending on Load Ratio. *Journal of Asian Architecture and Building Engineering,* 10(2), 2011.http://dx.doi.org/10.3130/jaabe.10.413.

16. Phan LT, McAllister TP, Gross JL, Hurley MJ. Best Practice Guidelines for Structural Fire Resistance Design of Concrete and Steel Buildings. Technical Note. Gaithersburg, Md: National Institute of Standards and Technology; 2010. Report No.: 1681 http://www.nist.gov/customcf/get_pdf.cfm?pub_id=907295.

17. Meléndez1 J, Foronda A, Aranda JM, López F, López de Cerro FJ. Infrared thermography of solid surfaces in a fire. *Measurement Science and Technology,* 21(10):105504,.http://dx.doi.org/10.1088/0957-0233/21/10/105504.

18. Inaudi D, Glisic B. Distributed Fiber Optic Strain and Temperature Sensing for Structural Health Monitoring. In The Third Int'l Conference on Bridge Maintenance, Safety and Management, IABMAS'06; 2006; Porto, Portugal.

19. Huang Y, Zhou Z, Zhang Y, Chen G, Hai X. A Temperature Self-Compensated LPFG Sensor for Large Strain Measurements at High Temperature. *IEEE Transactions on Instrumentation and Measurement,* 59(11):2997-3004, 2010.http://dx.doi.org/10.1109/TIM.2010.2047065.

20. Hausner MB, Suarez F, Glander KE, van de Giesen N, Selker JS, Tyler SW. Calibrating Single-Ended Fiber-Optic Raman Spectra Distributed

Temperature Sensing Data. *Sensors 2011,* 11(11):10859-10879, 2011.http://dx.doi.org/10.3390/s111110859.

21. Xiao H, Deng J, Pickrell G, May RG, Wang A. Single-Crystal Sapphire Fiber-Based Strain Sensor for High-Temperature Applications. *Journal of Lightwave Technology,* 21(10):2276, 2003.http://www.opticsinfobase.org/jlt/abstract.cfm?uri=jlt-21-10-2276.

22. Zhang B, Kahrizi M. High-Temperature Resistance Fiber Bragg Grating Temperature Sensor Fabrication. *IEEE Sensors Journal,* 7(4):586-591, 2007 April.http://dx.doi.org/10.1109/JSEN.2007.891941.

23. Bhatia V, Vengsarkar AM. Optical fiber long-period grating sensors. *Optics Letters,* 21(9):692-694, 1996.http://dx.doi.org/10.1364/OL.21.000692.

24. Vengsarkar AM, Lemaire PJ, Judkins JB, Bhatia V, Erdogan T, Sipe JE. Long-period fiber gratings as band-rejection filters. *Journal of Lightwave Technology,* 14(1):58-65, 1996 Jan.http://dx.doi.org/10.1109/50.476137.

25. Luna Technologies. Distributed Temperature and Strain Measurements. [Online].; 2012 [cited 2012 7 24. Available from: http://www.lunatechnologies.com/products/software/strain_temp_sensing.html.

26. LIOS Technology. Distributed Temperature Sensing. [Online].; 2012 [cited 2012 7 24 [1400 Campus Drive West, Morganville, New Jersey 07751]. Available from: http://www.lios-tech.com/Menu/Technology/Distributed+Temperature+Sensing.

27. Liu ZG, Ferrier G, Bao X, Zeng X, Yu Q, Kim AK. Brillouin scattering based distributed fiber optic temperature sensing for fire detection. In Seventh International Symposium on Fire Safety Conference; 2002; Worcester, MA USA. p. 221-232. http://www.nrc-cnrc.gc.ca/obj/irc/doc/pubs/nrcc45168/nrcc45168.pdf.

28. Dong Y, Zhang H, Chen L, Bao X. 2 cm spatial-resolution and 2 km range Brillouin optical fiber sensor using a transient differential pulse pair. *Applied Optics,* 51(9):1229-1235, 2012.http://dx.doi.org/10.1364/AO.51.001229.

29. Lonnermark A, Hedekvist PO, Ingason H. Gas temperature measurements using fibre Bragg grating during fire experiments in a tunnel. *Fire Safety Journal,* 43(2):119-126, 2008.http://dx.doi.org/10.1016/j.firesaf.2007.06.001.

30. Mentes Y, Kim YD, Zobel RS, Iadicola M, White DW, Leon RT, et al. Analytical and Experimental Assessment of Steel Truss Gusset Plate Connections. In Proceedings of International Bridge Conference; 2011;

Pittsburgh, PA.

31. Pan B, Wu D, Wang Z, Xia Y. High-temperature digital image correlation method for full-field deformation measurement at 1200 ºC. *Measurement Science and Technology,* 21(1):15701-15711, 2011.http://dx.doi.org/10.1088/0957-0233/22/1/015701.

32. Marr S. Underwater Visualization to Enhance Underwater Inspections. Marine Technology Reporter. 2010 October: p. 26-31.

33. National Aeronautics and Space Administration. NASA GRC Physical Sensors Instrumentation Research. [Online].; 2012 [cited 2012 7 24. Available from: http://www.grc.nasa.gov/WWW/sensors/PhySen/tech.htm.

34. Hitec Products. [Online].; 2012. Available from: http://www.hitecprod.com.

35. Touloukian YS, Kirby RK, Taylor RE, Desai PD. Thermal Expansion: Metallic Elements and Alloys: IFI/Plenum; 1975.

A   Results and Analysis of the Demonstration Test

A.1   Introduction and Motivation

The demonstration test had three goals.

- Evaluate the suitability of digital image correlation (DIC) for strain measurement under fire conditions.

- Compare digital image correlation to traditional high-temperature strain-gage measurements.

- Understand the sensitivity and uncertainty of both digital image correlation and high-temperature resistance strain gages in a real fire, as opposed to furnace, condition.

A.1.1   Background on Digital Image Correlation for High- Temperature Strain Measurement

Digital image correlation has been used for full-field strain measurement for more than twenty-five years. [1]. After several proof-of-concept demonstrations in the mid 1990s, [2, 3] recently researchers have been applying it to high-temperature strain measurement. [4–7] Most recent reports have also approached the problem as an experimental proof of concept, and are confined to methods for overcoming the significant problems of image distortion and illumination. They typically only report the successful measurement of thermal expansion strains in a test coupon in a furnace [4, 5, 7] or other bench-top environment. [6]

Digital image correlation is an image matching technique [8]. As implemented in most systems for measuring displacements and strains of solid bodies it relies on being able to compute the mapping (i.e. the correlation) of specific subset regions of the specimen, identified by their signature grey intensity level, between an initial and a deformed state. Uniquely identifying each region in an image through its grey level requires a random speckle pattern. Of course, if the specimen deforms, the coating must move with the surface to indicate strain in the specimen. It can be naturally occurring, etched into the surface, or applied as a coating. The preferred pattern has a minimum of three light-dark transitions in every direction within the subset of the image to be correlated. The displacements are determined in an average sense for each subset, so the subset size determines the footprint of a single independent displacement measurement. Reducing the subset size improves spatial resolution of the displacement field, but reduces the quality of the correlation. The correlation subset position is rastered across the images in the region of interest to determine a grid of displacements. As a rule of thumb, the minimum feature size of a pattern should be three pixels wide, resulting in a minimum correlation subset on nine pixels. In real applications, obtaining this idealized pattern for the entire surface is not possible, so patterns often have features larger than the three-pixel size and the subset for correlation is two to three times this minimum size.

Traditional two-dimensional digital-image correlation measures a displacement field on the surface of a flat specimen, and derives the strain field from this displacement field. All of the displacements and strains are based on the change from a reference image. The method uses a single camera to image a high contrast surface pattern on a flat specimen held at a fixed distance normal to the surface and at a fixed magnification. Out-of-plane motion must be minimized or compensated, since it will be seen as uniform expansion or contraction of the specimen. Out-of-plane bending produces a similar effect, but compensation is much more complex. Good quality images require a clear line of sight, low digital-image noise, and minimal shadowing, specular reflection, blurring, and degradation of the pattern, for example flaking of an applied coating.

Three-dimensional digital image correlation combines stereo-photogrametry and traditional DIC, and is used in the tests that this report describes. In this method, two digital cameras image the surface of interest from slightly different angles. After calibration of the orientation of the cameras, the stereo images of the surface can be used to see the surface pattern in three dimensions. This 3D pattern can then be correlated from the reference state to the current state to measure the 3D displacement field. From these displacements the surface strain field is calculated. Similar requirements on the pattern and image quality as mentioned above apply here, but three-dimensional surface shapes are now acceptable, as well as out of plane displacements. The only additional requirement is that both cameras must be able to image the pattern in the region to be measured. The pattern selection can be more complex, since the surface is not a a single effective magnification in either image even for a flat specimen, because each camera is angled to the surface normal, which results in a perspective distortion in each image.

Extending digital image correlation to high-temperature for strain measurements adds six challenges.

1. Convective air currents blur the image and refract individual rays of light. The latter can cause apparent specimen translation as well as apparent strain.

2. The index of refraction changes between the calibration phase and the test phase, and it varies with fire intensity and temperature.

3. Heating of DIC setup changes the camera orientation.

4. Flames licking the test specimen block the imaging, and general lighting changes from fire affect the measurement.

5. The fire assault may degrade the pattern, which may peel off, or soot build-up may obscure it.

6. Radiation from the specimen at high temperature may change the contrast of the image.

A.1.2 Background on High-Temperature Strain Gages

Resistance strain gages capable of operating at temperatures above 500 °C were developed in the 1960s, primarily to serve the aerospace industry. [9–11]. The basic operation and materials of the contemporary resistance strain gage are similar to those developed forty years ago. The field is sufficiently mature that ASTM International publishes a Standard Practice for using high-temperature strain gages. [12]

High-temperature strain gages suffer from some important limitations. Because their resistance depends on the microstructure of the metal that forms the element, oxidation and annealing of the wire can cause the gage resistance to change with time, producing apparent strain drift. This effect is generally a larger problem in configurations where strain measurement is made over days, rather the few hours envisioned for the measurements made in fire conditions. In addition, although the half-bridge configuration offers the possibility of self-compensation, some apparent strain as a function of temperature has proven to be unavoidable.

A.1.3 Outline

Section A.2 describes the evaluation and selection of suitable coatings for the digital image correlation measurements. Section A.3 describes the configuration and results of the demonstration test in the Large Fire Facility. Section A.4 describes several subsequent tests undertaken to understand the performance of the high-temperature strain gages.

A.2 Evaluation of Coatings For Digital Image Correlation

Digital image correlation requires that the area of interest be covered with a high-contrast, random pattern that produces regions with characteristic gray-level signatures, that the analysis software can track from image to image. For high-temperature digital image correlation, the black and white colors of the pattern must be stable against the temperature and fire. Other investigators have used different methods for producing the pattern, for example spattering the surface with different oxide pigments [5], or abrading the surface with SiC paper. [4].

Table 9: High-temperature paints evaluated

Coating	Paint	Manufacturer	Form
01	Thurmalox 230	Dampney Co.	liquid (white) aerosol (black)
02	Thermal Kote	Superior Ind.	aerosol
03	Zynolyte Hi-Temp Extreme Heat Enamel	Aervoe Ind.	aerosol

For this demonstration test, we chose to produce the pattern by painting the specimen with a matte white, high-temperature paint, and overspraying with a similar matte black high-temperature paint. These paints are sold commercially for painting wood stoves and automobile exhaust manifolds, as well as for protective coatings in the chemical process industry. We evaluated the performance of three types[*] of high-temperature paint, Table 9, using a coupon test. Both exposure temperature and surface preparation were evaluated. Exposure to open flame and the very high temperatures planned for the demonstration test are well outside the recommended use conditions of all three paints. Their performance in the demonstration test should not be construed as evaluation of their suitability for their *intended* use. All three coatings are proprietary formulations. The material safety data sheet (MSDS) for Coatings 01 and 02 specifically mention crystalline silica (CAS 14808-60-7 and CAS 7631-86-9 respectively). Coating 02 also contains aluminum flake (CAS 7429-09-5). The MSDS for Coating 03 lists only the hydrocarbon propellants.

A.2.1 Applying the Random Pattern for Digital Image Correlation

Digital-image correlation relies on the random, high-contrast pattern of light and dark regions that define, through their gray level, a region that can be located in each image. Based on the size of the test plate, the lenses used for imaging, the geometric arrangement of the cameras, and the potential for image blurring due to convective currents, a minimum black spot size of about 1.5 mm was determined for the demonstration test, Section A.3. All of the test coupons and the plates for the demonstration test were first painted with the white background over which black spots would be applied. We evaluated three methods for producing the black

---

[*]Certain commercial equipment, instruments, or materials are identified in this paper in order to specify the experimental procedure adequately. Such identification is not intended to imply recommendation or endorsement by the National Institute of Standards and Technology, nor is it intended to imply that the materials or equipment identified are necessarily the best available for the purpose.

portion of the high-contrast pattern with optimal spot size: partial spray, adhesive stencil, and particle masking. In the partial spray method, aerosol paint is over-sprayed onto the test piece. Larger droplets fall onto the test piece. This method produced black regions that were too small. A second method used an adhesive stencil with 2 mm diameter holes, but it produced patterns with insufficient density of black. This result, in addition to the difficulty making a stencil large enough for patterning a full-size plate, suggested a third patterning method. Since the plates being tested were nominally flat, a random distribution of non-adhering particles on the white paint coating could be used as a mask to prevent full coverage by the black paint. After removal of the particles, areas of the white base coat (slightly larger than the particles used) would remain on the plate. Seven types of particles ranging in aspect ratio from approximately 1:1 to 1:4 were tested; all had a smaller dimension < 2 mm. Sample patterning and evaluation suggested the aspect ratio of the best particle was near 1:1 and its diameter was between 0.5 mm and 1.0 mm. Although that particle size is smaller than the preferred feature size, the particles naturally clustered, and produced features very close to the desired 1.5 mm size. After the patterned plate was dry to the touch, the coating was cured for three heating and cooling cycles of 215 °C for about 40 minutes, followed by cooling to room temperature for about 60 minutes. Some yellowing of the white coating occurred during the first heating cycle, but most of this yellowing disappeared by the end of the curing process.

A.2.2 Furnace Coupon Test

Procedure The performance of the coatings was first evaluated in a furnace coupon test at different temperatures. The most promising coating was then tested to determine the robustness against direct contact with flame using a propane torch, and to evaluate the digital image correlation system in a more realistic scenario with flames and convection currents. Fifty-four mild steel coupons (50 mm × 25 mm × 6 mm) were prepared for a preliminary test of three brands of high-temperature paint and different surface preparation techniques. Coupons were engraved to indicate the factors tested. Codes were assigned according to brand of paint (01, 02, 03), surface Treatment (S, M, L), and test temperatures: (300, 400, 500, 600, 700, 800) °C.

The coupons were sandblasted using abrasive of three grit sizes: 100-170 (S), 60-100 (M), and 40-60 (L). The preparation instructions for Coating 2 recommended abrading the surface with a 60-100 sanding disk and curing the coating at 204 °C for 3 h before use. The other two coatings contained no surface preparation or pre-treating instructions, but all three coatings were prepared with the sandblasting and 204 °C cure.

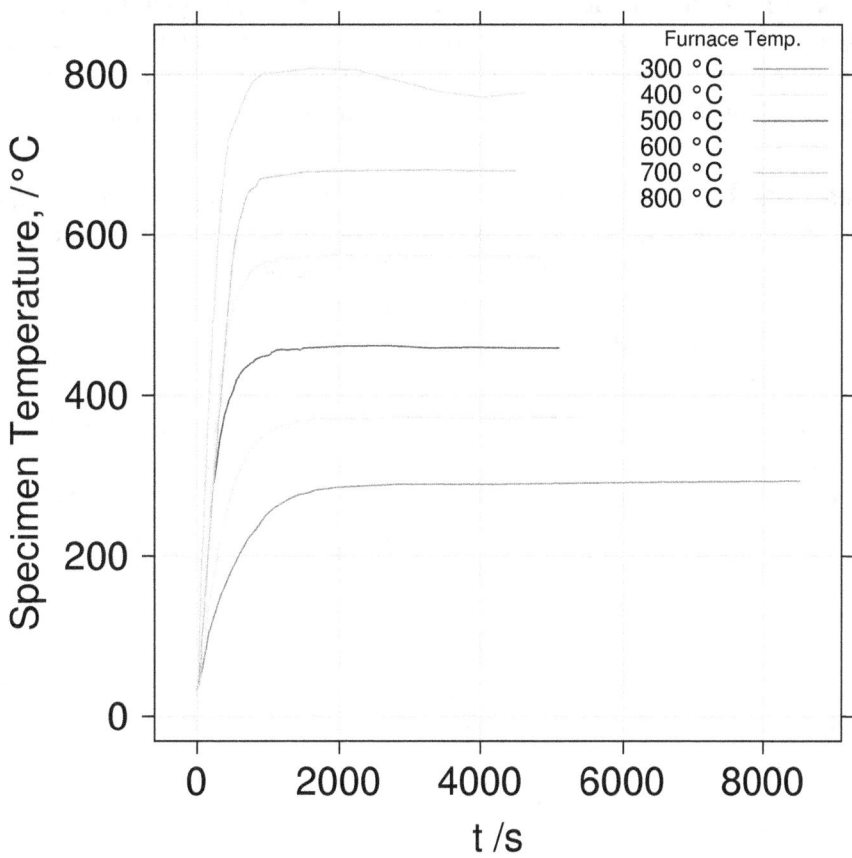

Figure 7: Temperature-time traces for the six coupon exposure tests.

To test the response of the paint to an elevated temperature, nine samples comprising three distinct surface treatments and three coating types were tested in air in a resistance-heated box furnace. The coupon temperature was monitored by peening a K-type thermocouple into a hole drilled to half the thickness of a dummy specimen. For the 300 °C and 800 °C tests the thermocouple was inserted in one of the actual test coupons. The furnace was set to the test temperature and allowed to stabilize before the test coupons, generally sitting on end in a rack, were inserted. For the tests with $T > 400$ °C, the coupons reached 95 % of the maximum temperature in less than 12 minutes. The two lower-temperature tests reached the 95 % level in 17 minutes and 26 minutes, respectively. Figure 7 shows the temperature-time traces of the specimens during the coupon tests. The coupons remained in the furnace for about one hour after their temperatures had stabilized, and were then removed and allowed to cool on the lab bench. Immediately after removal from the furnace, the coupons were photographed while still hot. They were also photographed again several hours later after cooling to room temperature.

Results For exposure temperatures less than 700 °C, the different surface preparations had no effect on the adherence of the coatings. Figure 8 is a montage of the results of five of the exposure tests, and shows before and after images. The images have been partially color corrected to facilitate comparison, but the images are not completely color consistent.

300 °C All the coupons were slightly discolored to a brownish hue, and Coating 02 developed a white speckle pattern in regions up to 1 cm in diameter. This report does not address the question whether this speckling is an inherent quality of the coating or if it represents a reaction with contaminants in the furnace, which had been used for ceramic research with volatile oxides for many years.

400 °C All coupons were moderately discolored, and the Coating 02 discoloration was more pronounced. In addition, the white speckle pattern on Coating 02 that developed in the 300 °C exposure became more pronounced. All discoloration and speckling occurred during the exposure. Unlike some of the higher temperature tests, the appearance of the coupons did not change between removal and cooling.

500 °C The results were identical to the 400 °C exposure.

600 °C Immediately after being removed from the furnace, the samples had a yellow tinge, possibly from the glow produced by the steel under the coat surface. All of the coatings adhered to the coupons, and there was no evidence of blistering or peeling. There was no speckling on the Coating 02 coupons. After cooling, all three coatings were white, and similar to the color before exposure. At room temperature, Coating 01 began to peel off the coupon in flakes up to 1 cm across. The other two coatings adhered to the coupons.

700 °C Immediately after removal, the samples had a orange tinge, possibly from the glow produced by the steel under the coat surface. All of the coatings adhered to the coupons, and there was no evidence of blistering or peeling. There was

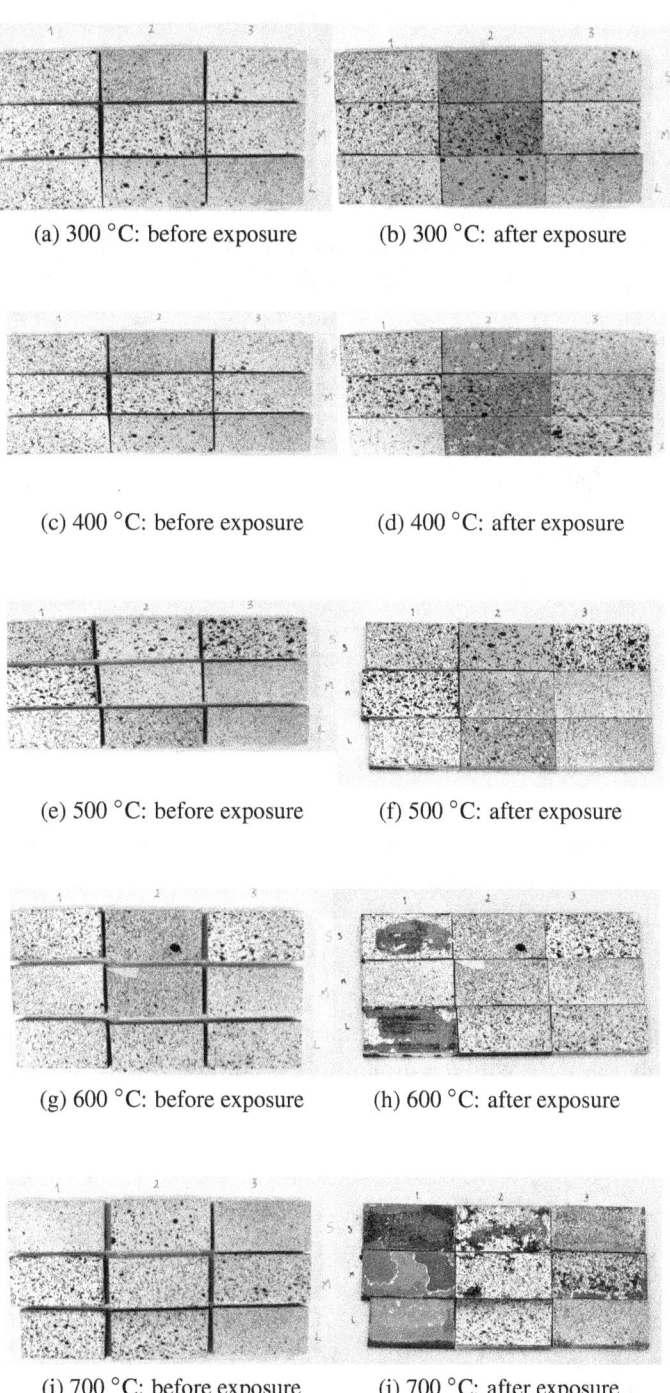

(a) 300 °C: before exposure   (b) 300 °C: after exposure

(c) 400 °C: before exposure   (d) 400 °C: after exposure

(e) 500 °C: before exposure   (f) 500 °C: after exposure

(g) 600 °C: before exposure   (h) 600 °C: after exposure

(i) 700 °C: before exposure   (j) 700 °C: after exposure

Figure 8: Images of the test coupons taken just before and several hours after the furnace test.

Table 10: Summary of coating exposure tests.

T °C	Coating 01	Coating 02	Coating 03
300	whiter tint	browner tint; 5 mm white speckling	whiter tint
400	whiter tint	browner tint; 5 mm white speckling	whiter tint
500	whiter tint	browner tint; 5 mm white speckling	whiter tint
600	peeled after cooling	adhered after cooling	adhered after cooling
700	all coatings flaked off	coatings mostly adhered	some coatings flaked off
800	all coatings flaked off	all coatings flaked off	all coatings flaked off

no speckling on the Coating 02 coupons. After cooling, the paint on all samples began to peel. Coating 01 peeled the most severely. Handling dislodged more of the coating, and all of Coating 01 fell off. Coating 02L and Coating 03L (sandblasted with the coarsest grit) adhered the most.

800 °C None of the coupons retained any coating. The coupons were oxidized under the coatings. In this test, the coupons had been supported upright by wires which in some cases had touched the coatings. In these areas the oxidation and coating failure was more pronounced.

Conclusions from exposure tests Table 10 summarizes the results of the furnace exposure tests. None of the coatings performed well at 700 °C or above. Because it adhered well and did not discolor or develop the white speckling, which would certainly interfere with the digital image correlation, the demonstration test used Coating 03.

A.2.3 Flame-lick Test

To gain experience prior to the full-scale demonstration test, we evaluated the performance of Coating 03, with two different spot patterns, as well as the performance of the digital image correlation cameras and filters, in bench-scale tests. In this test, two propane torches heated a thermocouple-instrumented coupon. The K-type thermocouple was peened into a hole drilled to half the depth of the thickness on the backside of the coupon. The tip of the flame of each torch projected into the field of view of the cameras, and directly hit the coated surface in the region of the thermocouple. Within ten minutes, the temperature of the coupon exceeded 700 °C. Because the cameras were unfiltered, the infra-red radiation from the coupon overwhelmed the auto-level capability of the camera. To the naked eye, the black spot pattern was clearly visible, but the camera imaged only a white center region. Presumably the proper filtering used in the phase 3 of the demonstration test would remove this problem. Although the test lasted for about fifteen minutes, the coating did not flake off either during or after the test. This behavior suggests that it was the formation of the oxide layer, rather than the exposure to temperature, that caused the coatings to delaminate in the furnace exposure tests.

A.3   Demonstration Test

The demonstration test took place in the National Fire Research Laboratory on 2011-09-15 in three phases. The test evaluated the displacement and strain resolution of the digital image correlation system by imposing known rigid-body translations of the plate and comparing the computed in-plane strains, which should remain identically zero. In phase 1, the plate was translated vertically and horizontally known amounts to determine the sensitivity of the digital image correlation system with no heat applied. In phase 2, the burner was placed about 600 mm in front of the plate, to test the effect of convection with minimal heating to the plate. In phase 3, the plate was suspended by an arm directly over the natural gas burner to test the combined effects of convection, heating, and flame lick. Figure 9 shows the general layout of the test, with the plate suspended above the gas burner during phase 3. Figure 10 is a schematic diagram of the test configuration, viewed from above, that shows the relationship of the plate, the burner, the digital image correlation camera system, and the coordinate system.   During phase 3 of the test, with the burner directly under the plate, the heat output, P, was in the range $(0 < P < 70)$ kW with specific holds at 30 kW, 40 kW, 50 kW, and 70 kW.

A.3.1   Test Configuration

The original experimental plan was to use a natural gas fire to heat a instrumented mild-steel plate, ultimately to realistic fire temperatures. The 30 cm x 45 cm x 1 cm plate is instrumented with eight type K thermocouples that are peened into holes drilled roughly half the depth of the plate. Thermocouples 2 and 3 are mounted adjacent to gage 1, and thermocouple 7 is mounted adjacent to gage 2. Thermocouple 8 broke off during installation. The front side of the plate is painted with a special high-temperature paint to produce a random black-and-white pattern to enable strain measurement by digital image correlation. The back side of the plate

Figure 9: Experimental configuration during the Phase 3 of the demonstration test, with the burner in position 2.

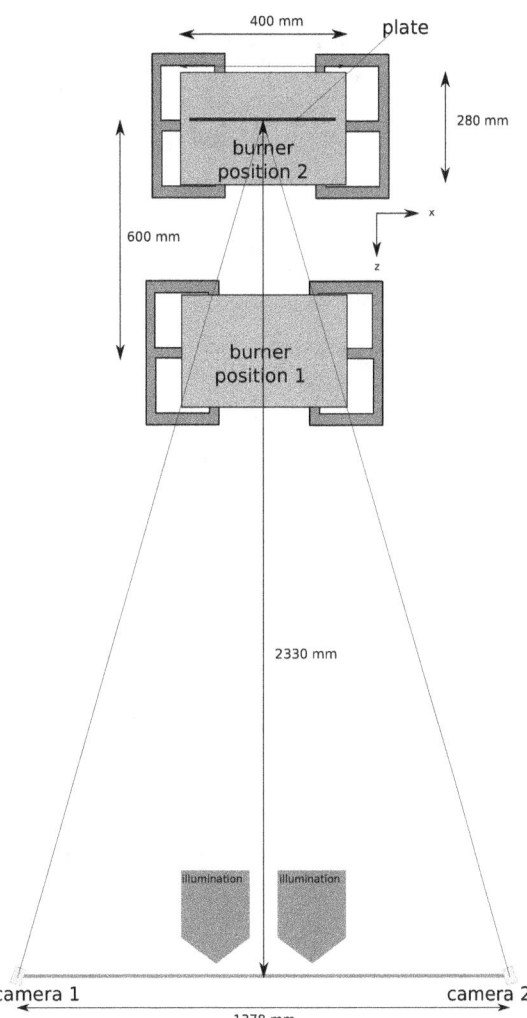

Figure 10: Layout of the digital image correlation camera system and instrumented plate. View from the top.

52

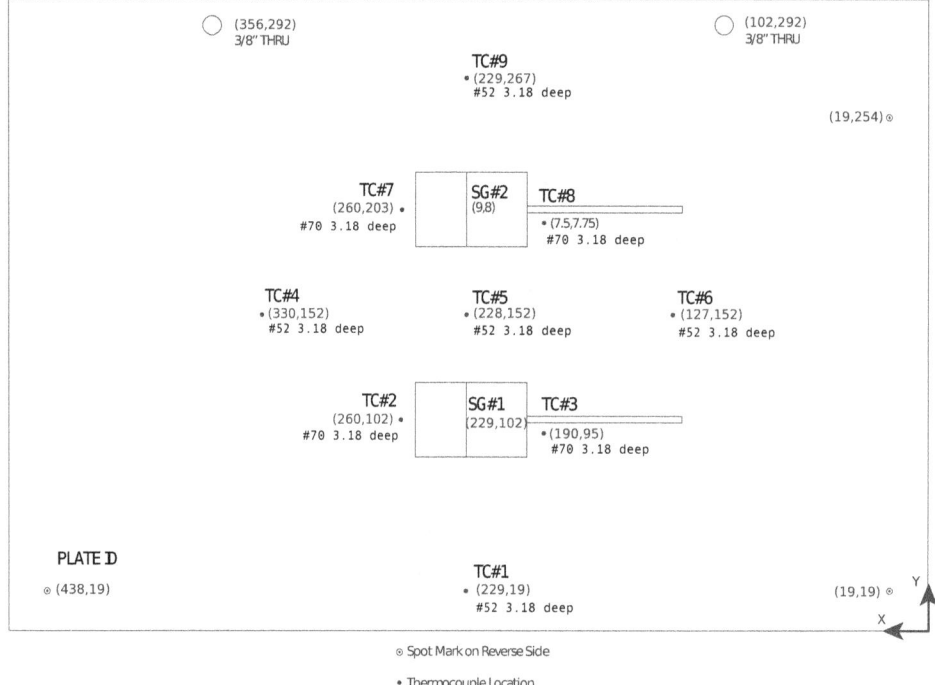

Figure 11: Location of gages and thermocouples on the plate, as viewed from the back side.

is instrumented with the eight thermocouples and two strain gages, which were installed by their manufacturer. Figure 11 shows the locations of the gages and the thermocouples, viewed from the back side of the plate. The reference marks shown in Figure 11 at 19 mm from the plate edge appear in the pattern on the front side of the plate. These marks are used to rotate the DIC results to the plate coordinate system shown in Figure 11.

A.3.2 Digital Image Correlation Setup

Figure 10 shows the orientation of the cameras to the plate for the demonstration test , where the plate and burner are under the fume hood and the cameras and illumination are just outside of the hood. Two 5 MP Flea 2 CCD cameras are mounted to an aluminum extrusion bar and are angled approximately ± 16.5 degrees to the surface normal, resulting in a 33 degree included angle. The mounting bar is about 2330 mm from the face of the plate at a height that places the cameras' optical centerlines at the initial centerline of the plate. Each individual image is recorded as a 2448 pixel x 2048 pixel 8-bit gray-scale tagged image file (TIF). Both cameras

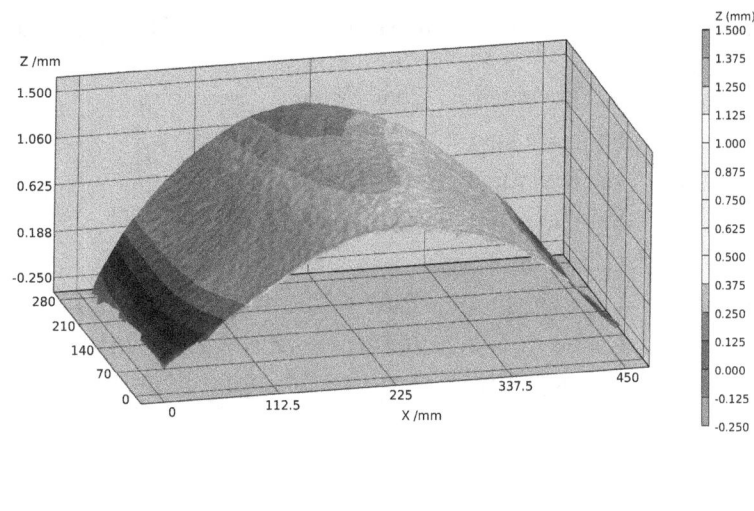

Figure 12: Shape of the plate as measured by digital image correlation.

use Schneider 35 mm F1.9 C-mount Xenoplan compact lenses with locking focus. Each lens was fitted with a UV-IR cut off filter (IF486). These filters pass more than 90 % of the visible light in the range of $(390 < \lambda < 690)$ nm. Based on this setup, the average effective magnification of the plate was 0.243 mm/pixel. Two 500 W incandescent photographic lights illuminated the plate, with the intention to overpower the potential light from the flame. Using this lighting and an aperture of f/7 on the lenses, an exposure time of 15 ms was needed to achieve a good-quality high-contrast image.

After initial set up, the camera, lenses, and their orientation to one another was calibrated using 75 image pairs of a calibration grid of points (spaced at 38.1 mm). These images were processed using the VIC-3D 2007 software from Correlated Solutions Inc. An initial image pair of the hanging plate was processed using the VIC-3D software and the calibration that was developed. The DIC 3D plot in Figure 12 shows that the plate is slightly curved. It also illustrates the sensitivity of the digital image correlation measurement to changes in shape.

Image acquisition during the demonstration test was performed using VIC Snap 2007 software from Correlated Solutions Inc. During phases 1 and 2 of the demonstration test, at each plate position, approximately 60 sequential image pairs were acquired at intervals of 200 ms. During the phase 3 test, the reference image pair was taken during phase 1, before the fire was lit. Then image pairs were acquired at 200 ms intervals until the fire was extinguished, after which time image pairs were taken at 30 s intervals for another 4400 s. Post-processing of the image

Table 11: Characteristics of the HBWAH-12-250-6-2CB-HB weldable strain gage

Parameter	Code	Value
Shim thickness		0.13 mm (0.005 in)
Gage resistance	12	120 Ω
Shim material	6	Hastelloy X
Gage factor, G		2.42
Gage length	250	6.3 mm (0.25 in)
Cable	2CB	braided Nextel 2 ft (610 mm) long
Extension wires	HB	Hoskins alloy 875

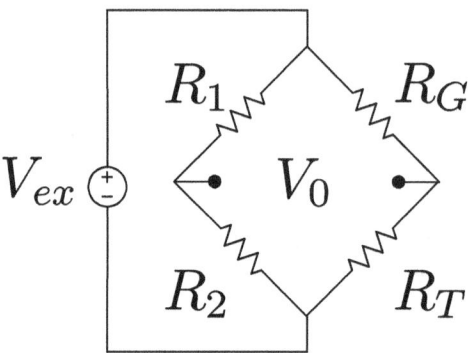

Figure 13: Bridge configuration

pairs used a subset of 31 pixels (approximately 7.5 mm) with a raster step size of 15 pixels (approximately 3.6 mm) resulting in about 9000 node points across the face of the plate. These parameters were selected based on experience, and were not optimized.

A.3.3 Strain Gages

The two strain gages were HITEC model HBWAH-12-250-6-2CB-HB gages. Table 11 summarizes the characteristics of the gages. These gages have an active gage with $R = 120$ Ω attached to a Hastelloy X shim that is in turn spot welded to the plate. A second compensation gage wired as the other leg in the Wheatstone bridge, R_T in Figure 13, is attached to a mild-steel "compensation block" that the vendor supplied. The compensation block is welded to the shim at only one end, so it expands freely with temperature. In this way, the thermal expansion of the plate is compensated or nulled. If the temperature of the active and compensation gages

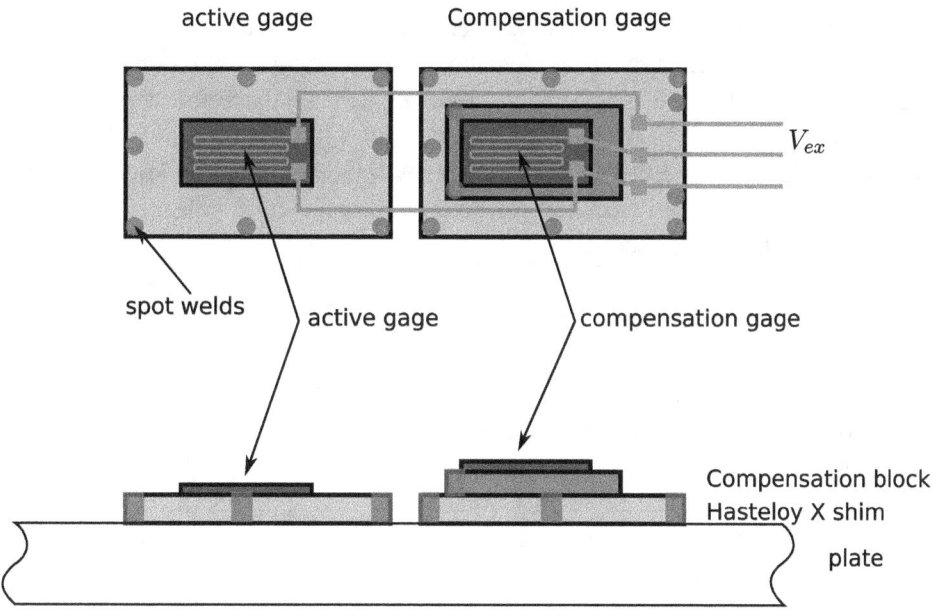

Figure 14: Schematic diagram of the strain gages used in the demonstration test.

changes the same amount, the changes in their resistance should also be nulled, and produce no change in the output of the gage. In a non-isothermal test, it is possible that the active and compensating gages will not be at the same temperature, and the gage may register an apparent strain. The entire gage assembly is protected by a stainless steel cover also welded to the plate. Figure 14 is a schematic diagram of the structure of the gages, redrawn from the manufacturer's literature. During both tests that this report describes, the two gages were conditioned by Measurements Group, Inc model 2210 signal conditioners, with 5.00 V excitation and 10x gain of the resulting signal.

A.3.4   Strain Calculation from Gage Output

Figure 13 shows the generic half-bridge strain gage circuit. In the figure, R_G is the active gage, and R_T is the temperature compensating gage. The ratio of the output to input voltages, ignoring any gain, A, applied to the output in this circuit, is

$$\frac{V_0}{V_{ex}} = \frac{R_T}{R_T + R_G} - \frac{R_2}{R_1 + R_2} \tag{2}$$

The change in resistance with strain, ϵ, in a gage is defined as

$$\Delta R_x = G R_x \epsilon \tag{3}$$

where the proportionality constant G, is termed the gage factor.

The compensating gage, R_T is meant to correct or compensate for temperature-induced changes in the resistance of the strain gage material and for thermal strains in the material being measured, in this case the plate. If the strains in the active and compensating gages are different, the bridge will have a non-zero voltage output at V_0. Assume that the resistances of the bridge-completion resistors are equal: $R_1 = R_2$. Then Eq. (2) becomes

$$\frac{V_0}{V_{ex}} = \frac{1 + G\epsilon_T}{2 + G(\epsilon_T + \epsilon_G)} - \frac{1}{2} \tag{4}$$

If the strains in the active ("G") and compensating ("T") elements are equal, the bridge output remains zero. If the strain in the active gage ("G") is greater than the strain in the compensating ("T") gage, the first term is less than 1/2, and the resulting output is negative. To first order, if the temperatures of the active and compensating gages are identical, this configuration nulls the temperature-induced change in resistance as well as the thermal expansion strain of the plate.

A.3.5 Demonstration Test Results

The results of each of the three phases of testing are described below in regards to the specific effects of interest in each phase. Some comparisons are made across different phases of testing where appropriate.

Effect of rigid body motion in the absence of heating or turbulence   In phase 1 of the demonstration test, the plate was moved to nine different positions and the resulting apparent plate strains, which should remain identically zero, were computed by digital image correlation. This set of results provides a baseline on the uncertainty in the measurement in the absence of both heating and distortion of the images caused by the turbulent atmospheric convection caused by the fire. Figure 15 shows the computed e_{xx} strain across the surface of the plate for an image from phase 1 of the test. The contours of strains are superimposed on the actual image of the plate. The strains, which should be identically zero, lie in the range ± 150 μm/m.

Figure 16 shows the computed e_{xx} and e_{yy} strains evaluated at a single position near strain gage 1 for nine different plate positions in space. The path of the plate, computed from the digital image correlation image data is shown in Fig. 16a. Although the individual determinations of e_{xx} and e_{yy} are statistically distinguishable, the mean values range over less than 50 μm/m. In perspective, this uncertainty corresponds to the elastic strain that a 10 MPa stress produces, or to the thermal strain that a 4 °C temperature change produces.

Figure 15: Computed e_{xx} strain across the plate during phase 1.

Effect of convection on measurements by digital image correlation In phase 2 of the demonstration test, the burner was in position 1, between the digital image correlation camera system and the plate, see Fig. 10. From these results it is possible to asses the effects of the turbulent column of air rising from the fire, in the absence of significant heating to the plate. During this test segment the plate was moved to nine different positions in a nominal x-y plane relative to the focal plane of the camera system. The z coordinate is nominally out of the plane defined by the plate. Since the translation of the plate amounts to a rigid-body motion, the digital image correlation system should not measure any strain in the plate.

Figure 16d shows the computed strains in the x (horizontal) and y (vertical) directions on the plate, evaluated at a point directly opposite strain gage 1. Each determination at a given position consists of approximately 60 images, spaced nominally 0.2 s apart. The solid line is a moving regression to the entire data set. The scatter in the data is 150 times as large as in phase 1, with no fire and minimal convection, Figure 16b and c, but the mean value at each position is not significantly different from zero in all but two of the eighteen cases.

The two data sets depicted in Figure 17 illustrate the effect of the convection on the measurement of strain and displacement in the absence of plate motion. Figure 17 compares the x and y strains and the U, V, W displacements (corre-

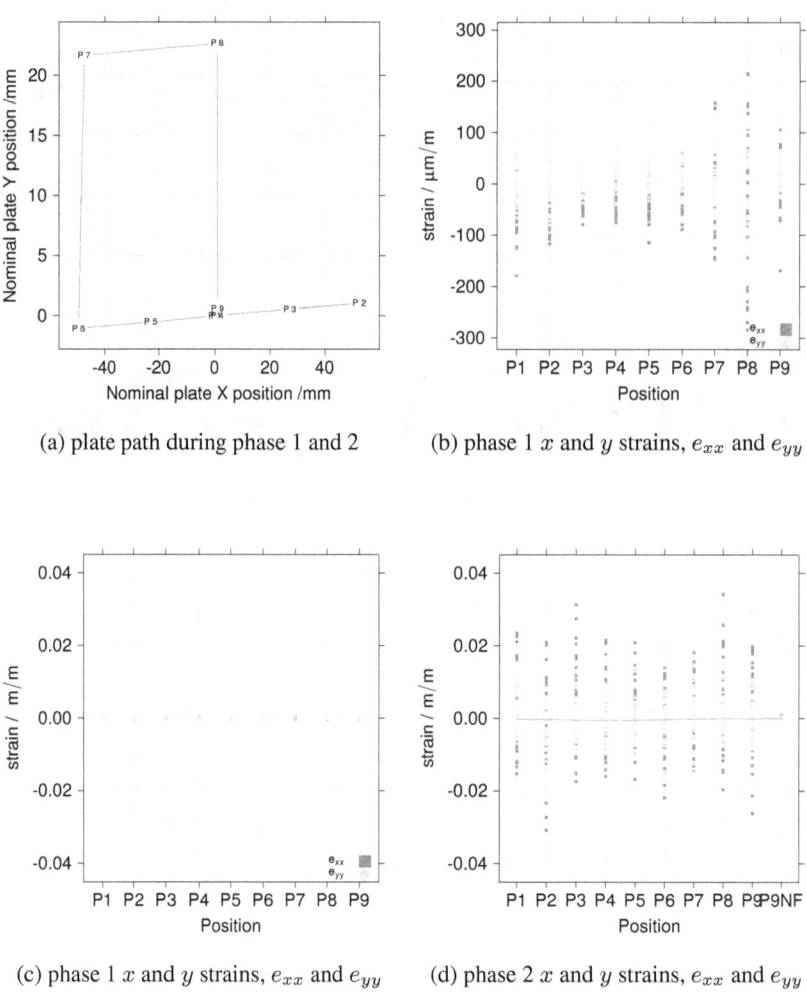

(a) plate path during phase 1 and 2

(b) phase 1 x and y strains, e_{xx} and e_{yy}

(c) phase 1 x and y strains, e_{xx} and e_{yy}

(d) phase 2 x and y strains, e_{xx} and e_{yy}

Figure 16: DIC-computed strains during phase 1 and phase 2. (a) path of the plate in space during the tests; (b) x and y strains in phase 1; (c) x and y strains in phase 1, same scale as in (d); (d) x and y strains in phase 2.

sponding to the x, y, and z directions relative to the plate) for the fire on and off conditions. The scales for the vertical axes for the strains and the positions are identical. Although the scatter in the strains e_{xx} and e_{yy} are much larger with the fire, and presumably with convection and resulting image distortion present, the mean values are indistinguishable from zero. Figures 17c-e compare the DIC-determined position of strain gage 1 for the two conditions, on identical scales. The U and V positions, corresponding to the x and y position in the plane of the plate, are unchanged, though the presence of the fire increases the scatter. The W position, essentially the position in the direction normal to the plane of the plate changes between the two conditions. The most significant result of this phase of the demonstration test is the invariance of the computed average in-plane strains e_{xx} and e_{yy} to the presence of the convective currents in front of the plate.

Effect of direct heating of the plate Figure 18 shows the output of the eight thermocouples on the plate during phase 3, in which the burner heated the plate directly. The plate reached a maximum temperature of about 200 °C during the test. After the peak temperature was reached, the fire was extinguished and the plate cooled naturally back to room temperature. Figure 19 shows the calculated thermal expansion strain near the two gages (dashed lines). The strains were calculated from a polynomial expression

$$\frac{\Delta L}{L_0} = l_0 + l_1 T + l_2 T^2 + l_3 T^3 \tag{5}$$

The data for this expression, Table 12, come from the curve labeled "Provisional" in Figure 261 of Touloukian's [13] compendium of thermal expansion of metals and alloys. Table 13 summarizes the regression of Eq. (5) on the data. Figure 19 also shows the strains (solid lines) computed from the output of the two strain gages, using Eq. (4). The gages are designed to correct for thermal expansion strains and for apparent strain caused by the change in temperature of the gage, so their EMF should not change. Instead they each read an apparent compressive strain.

Figure 20 plots the strains reported from the two strain gages, computed from Eq. (4), the strains computed from the digital image correlation image data in the region of gages 1 and 2, and and the thermal strains computed from the thermal expansion data, Eq. (5). during a 100 s window near the time of the peak plate temperature. The lines for the DIC-computed strains are a moving fit to the individual values for each image, since the digital image correlation strains and the temperature and strain-gage data are not on common time bases. The agreement between the digital image correlation data and the predicted thermal expansion strain is striking. Figure 21 plots the measured strains in the horizontal and vertical direc-

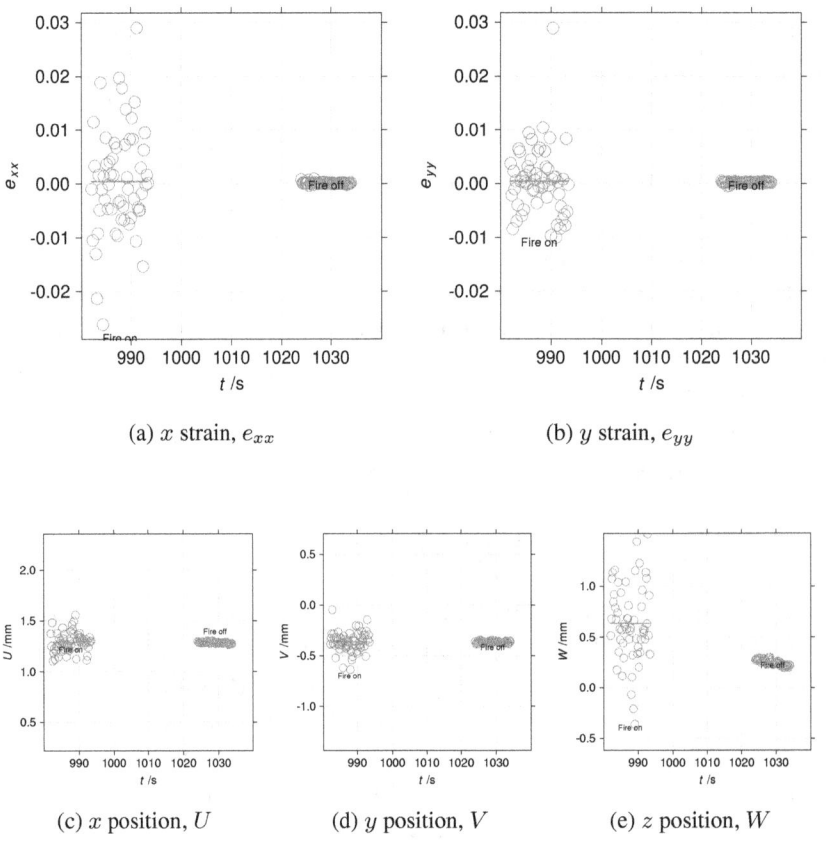

(a) x strain, e_{xx}

(b) y strain, e_{yy}

(c) x position, U

(d) y position, V

(e) z position, W

Figure 17: DIC-computed strains and positions during the portion of the test with the fire in burner position 1 ("between") showing effect of convection "fire on" $(980 < t < 995)$ s vs. "fire off" $(1025 < t < 1035)$ s.

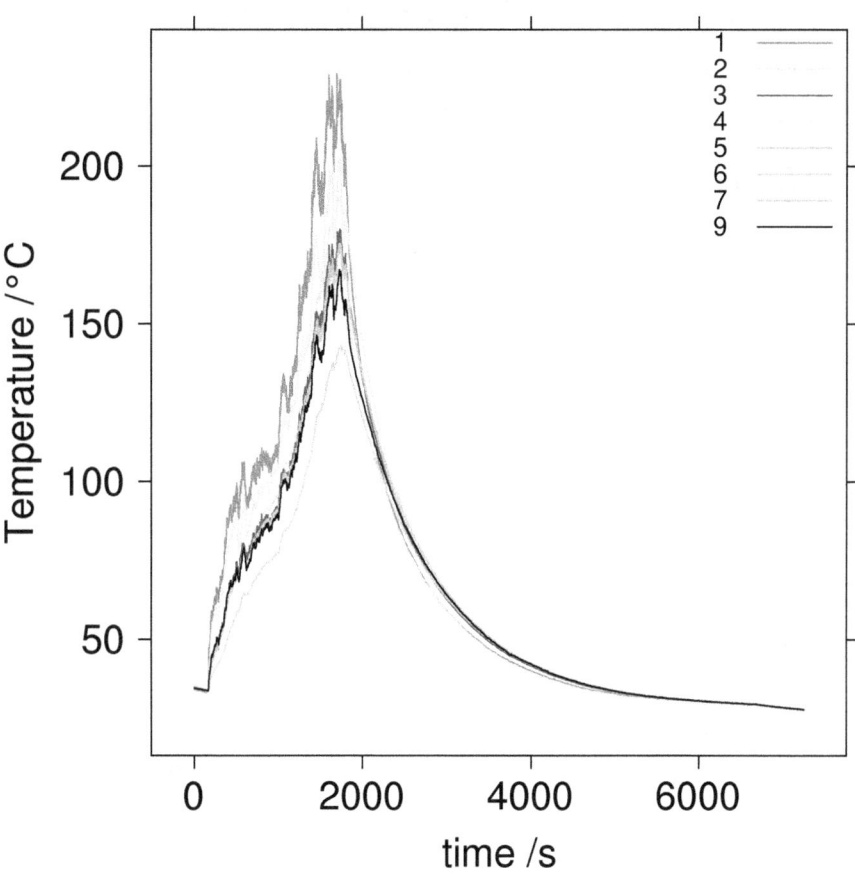

Figure 18: Temperature of the plate during phase 3 of the demonstration test.

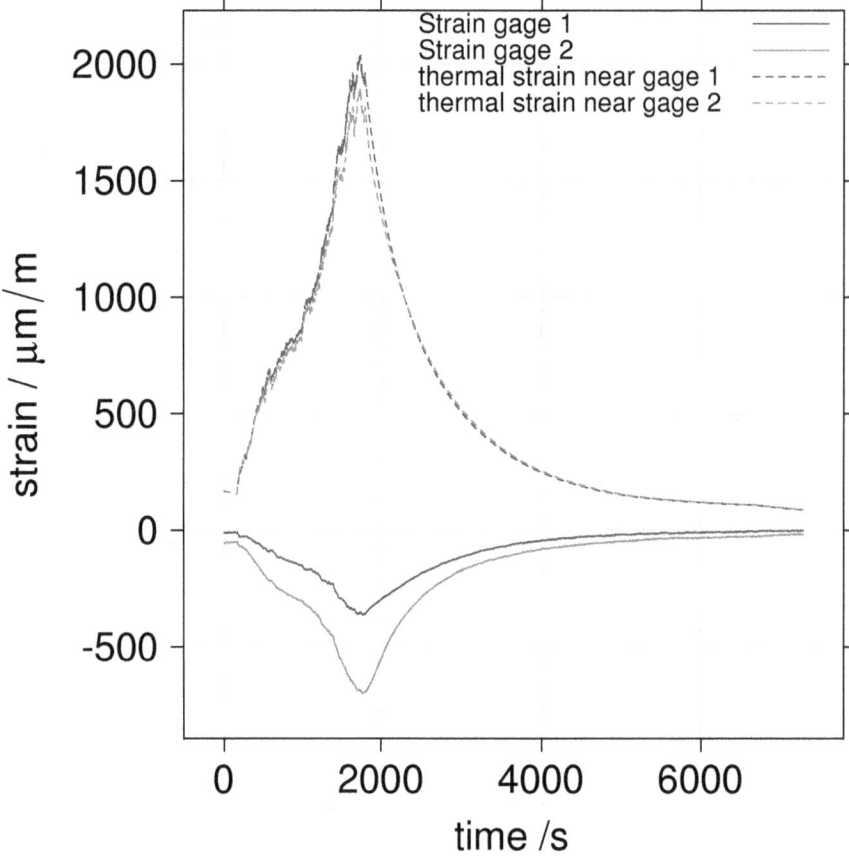

Figure 19: Computed thermal strain, , Eq. (5), and computed strain reported by the two strain gages during the demonstration test.

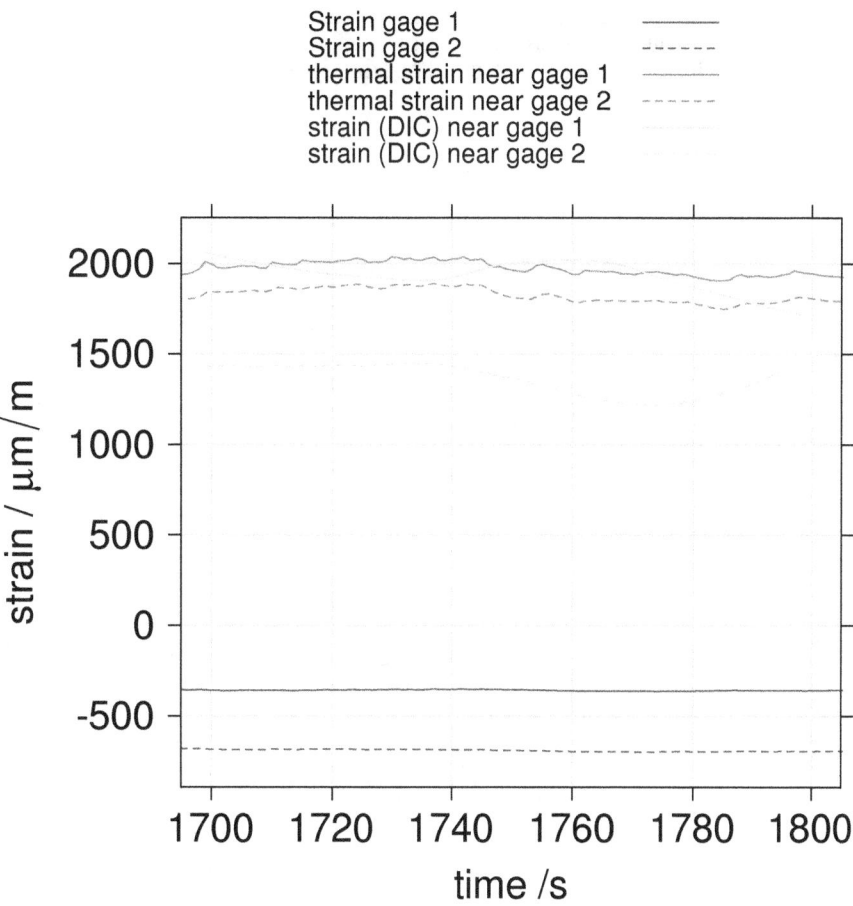

Figure 20: Comparison of the $x - x$ strains measured by DIC and the two strain gages with the thermal expansion strain computed from Eq.(5) during the time of the peak temperature.

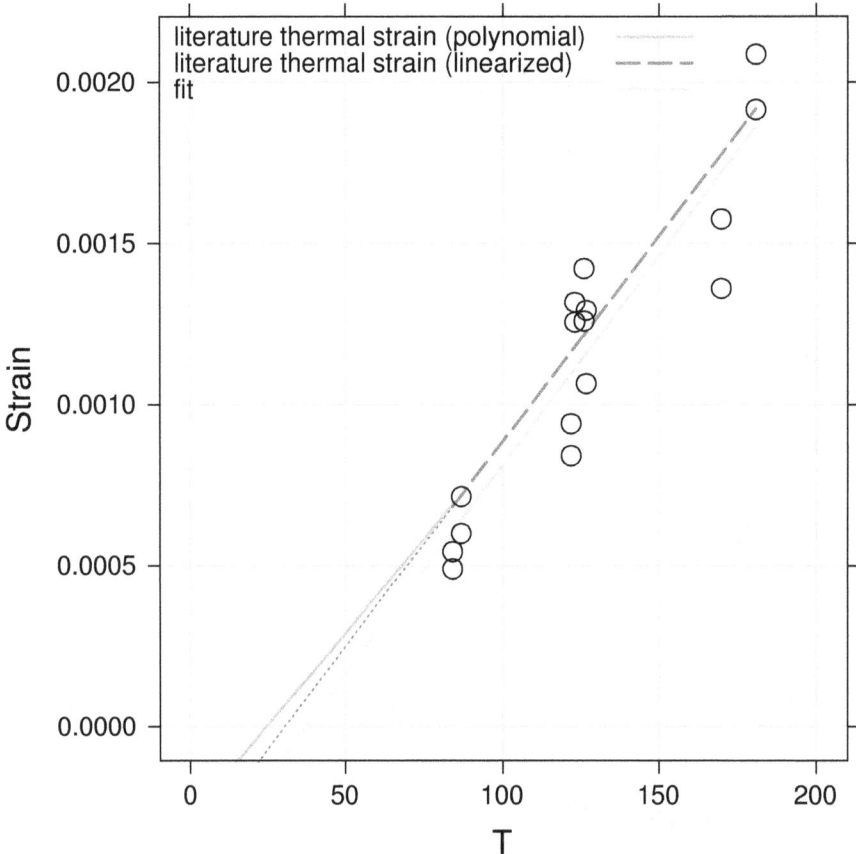

Figure 21: Thermal strain measured by DIC as a function of temperature in the regions of gages 1 and 2. The solid gray line is the thermal expansion strain computed from Eq. (5). Data for phase three of the demonstration test. In the region of the data, that polynomial expression has been linearized (dashed line). The green dash/dot line is the linear regression of the data.

Table 12: Data used to estimate the parameters of Eq. 5

| T | $\Delta L/L_0$ |
$^\circ$C	
201	-0.0010339
265	-0.0003330
333	0.0003679
393	0.0010796
467	0.0020976
530	0.0029733
587	0.0037505
656	0.0048449
735	0.0060161
809	0.0072417
883	0.0083909

Table 13: Summary of the non-linear regression to estimate the parameters of Eq. 5

Parameter	Estimate	Standard Error	Units	t
l_0	-2.4661×10^{-3}	1.5289×10^{-4}		-16.13
l_1	4.8662×10^{-6}	1.0132×10^{-6}	$^\circ$C^{-1}	4.80
l_2	1.2640×10^{-8}	2.0187×10^{-9}	$^\circ$C^{-2}	6.26
l_3	-4.7712×10^{-12}	1.2353×10^{-12}	$^\circ$C^{-3}	-3.86

RSD: 3.046×10^{-5} on 7 degrees of freedom

tions measured from the digital image correlation system in the region of gages 1 and 2 for two additional 100 s windows during the heating portion of phase 2, as well as for one portion during the cooling portion, with the burner off. The solid gray line is the thermal expansion strain computed from Eq. (5) referenced to the temperature, $T = 24.6\,^\circ$C . The dashed line is a linearization of that polynomial expression in the range of the data. The dash/dot line is the fit to the data. The slopes of the lines in the region of the data are quite similar: $\alpha_{\text{fit}} = 1.30 \times 10^{-5}$ and $\alpha_{\text{lit}} = 1.28 \times 10^{-5}$, a difference of less than 1.9 %In addition, the fit to the strain data differs from the literature value by less than 11 % everywhere in the range of the data. . Table 14 summarizes the data plotted in Figure 21. The temperature uncertainties in Table 14 represent the range of the temperatures during the 100 s window. The strain uncertainties are the usual standard uncertainty of the

Table 14: Comparison of mean strains, $\bar{\epsilon}$ measured by digital image correlation and computed from thermal expansion, Eq. (5), during four time increments during phase 3 of the demonstration test. Data are plotted in Fig.21

Gage	T	$\bar{e}_{xx}$	$\bar{e}_{yy}$	$\Delta L/L_0$
	°C	μm/m	μm/m	μm/m
1	126 ± 4	1258 ± 79	1422 ± 87	1208
1	87 ± 1	599 ± 772	713 ± 634	724
1	127 ± 2	1064 ± 1493	1292 ± 977	1217
1	181 ± 3	1915 ± 1628	2086 ± 1504	1926
2	123 ± 3	1255 ± 72	1317 ± 72	1171
2	84 ± 1	489 ± 706	543 ± 618	691
2	122 ± 2	840 ± 863	940 ± 951	1156
2	170 ± 3	1360 ± 1344	1575 ± 974	1778

mean value. In computing the mean values, strain points whose value was more than two standard deviations from the mean value were excluded. These generally came from images in which a flame lick appears. In general, not more than 10 % of the images were excluded. The uncertainty in the data for temperatures about 125 °C originate from a portion in the cooling phase, with no flame present. To within the accuracy of the original expression for thermal expansion strain, the digital image correlation system has measured the thermal expansion strains of the plate.

A.4   Subsequent Gage Verification Tests

The strain gages were not fully temperature compensated during phase 3, see Figure 19, and produced apparent compressive strains on heating. Two other tests provided information about the source of this discrepancy: a mechanical verification test at room temperature, and an isothermal heating test in an oven.

A.4.1   Gage Verification Test

After the demonstration test, we checked the proper room-temperature operation of the two strain gages by verifying against the strain measured in the plate using digital image correlation. We elastically bent the plate, while it was still suspended in the test fixture, and measured the output of the gages on the compression side and simultaneously measuring the elastic strain near the gage using the digital image

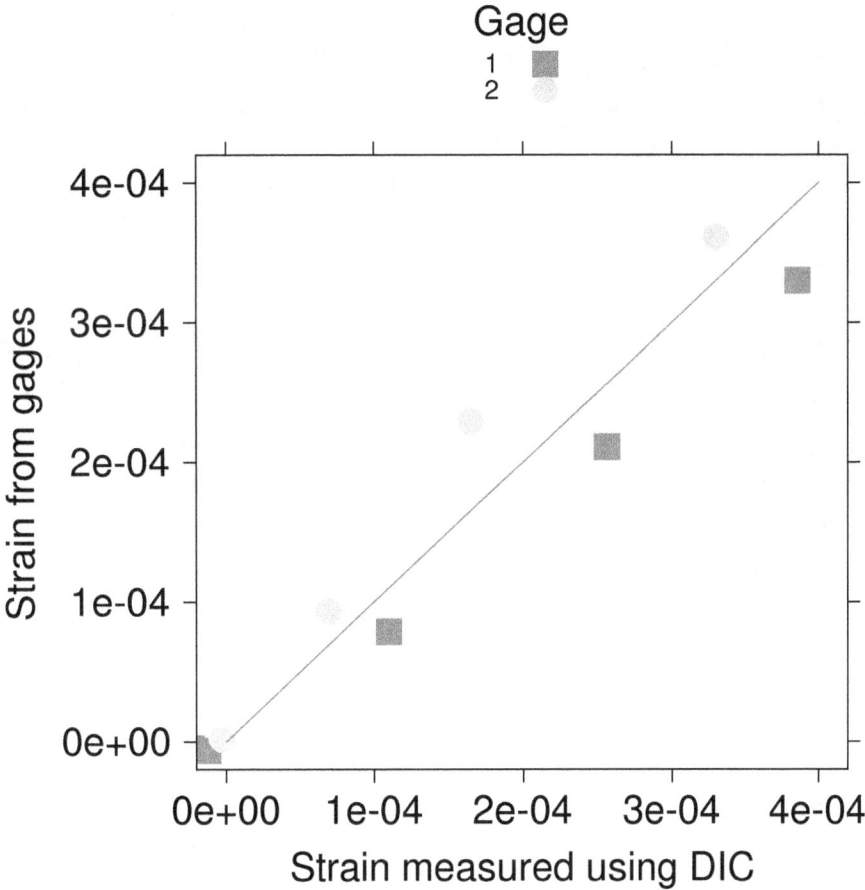

Figure 22: Strains computed from the gage output compared to strains measured using digital image correlation during elastic bending of the plate.

correlation setup.

Figure 22 shows the result of that test. The sign of the strain from the gages has been reversed, since the gages were on the compression side of the plate, and the DIC-measured strains were on the tensile side of the plate. The slope of the solid line is unity. Given the simple bending fixture and all the approximations, the agreement is excellent. The results of this experiment show that the gages are operating as expected for mechanical deformations at room temperature.

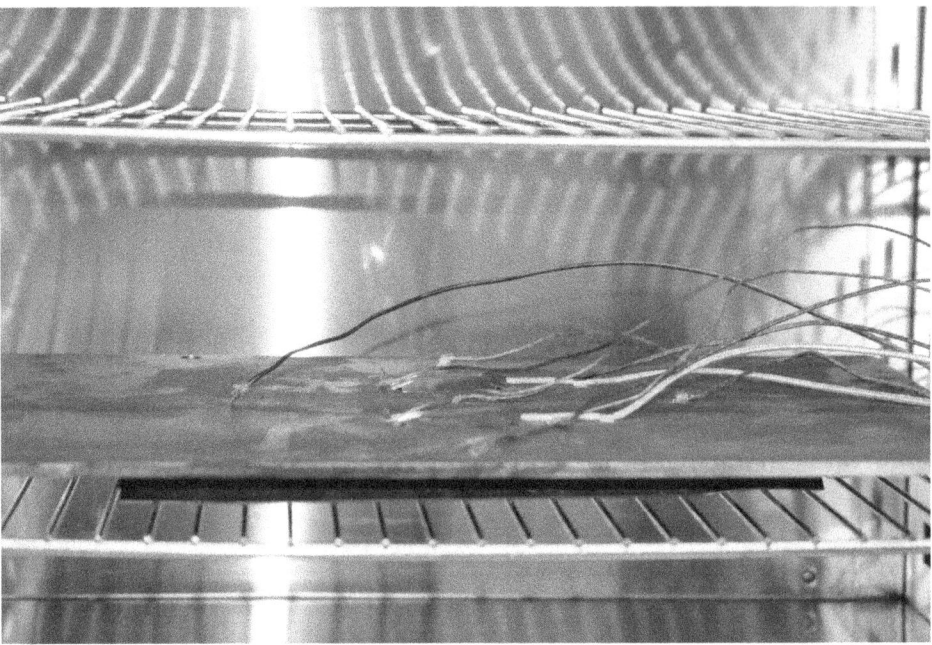

Figure 23: The instrumented plate in the convection oven.

A.4.2 Oven Test

To test the possible contribution of the non-isothermal temperature field on the output of the strain gages, we conducted a heating and cooling test on the instrumented plate in a laboratory convection oven with isothermal holds at 150 °C, 200 °C, and 250 °C. Figure 23 shows the plate mounted in the convection oven.

Figure 24 summarizes the results of the oven test to determine the response of the strain gages to temperature. Figure 24a shows the plate temperature during the oven test. The third, fourth, and fifth holds could have been longer, but the plate temperature was constant to within a degree during those holds as well. The leading curve labeled "monitor" corresponds to a thermocouple mounted in the airspace above the plate. Figure 24b shows the output of gage 2 during the oven test. The negative offset at each temperature is reproducible and depends linearly on temperature, see Figure 24c. Figure 24d shows the voltage output data of Figure 24b transformed to strain, after including the $A = 10$ gain of the output signal.

The linearity and reproducibility of the gage voltage with temperature, Figure 24c, demonstrate that the apparent strain from the gage is an inherent limitation of the gage and is not driven by differences in temperature between the active and

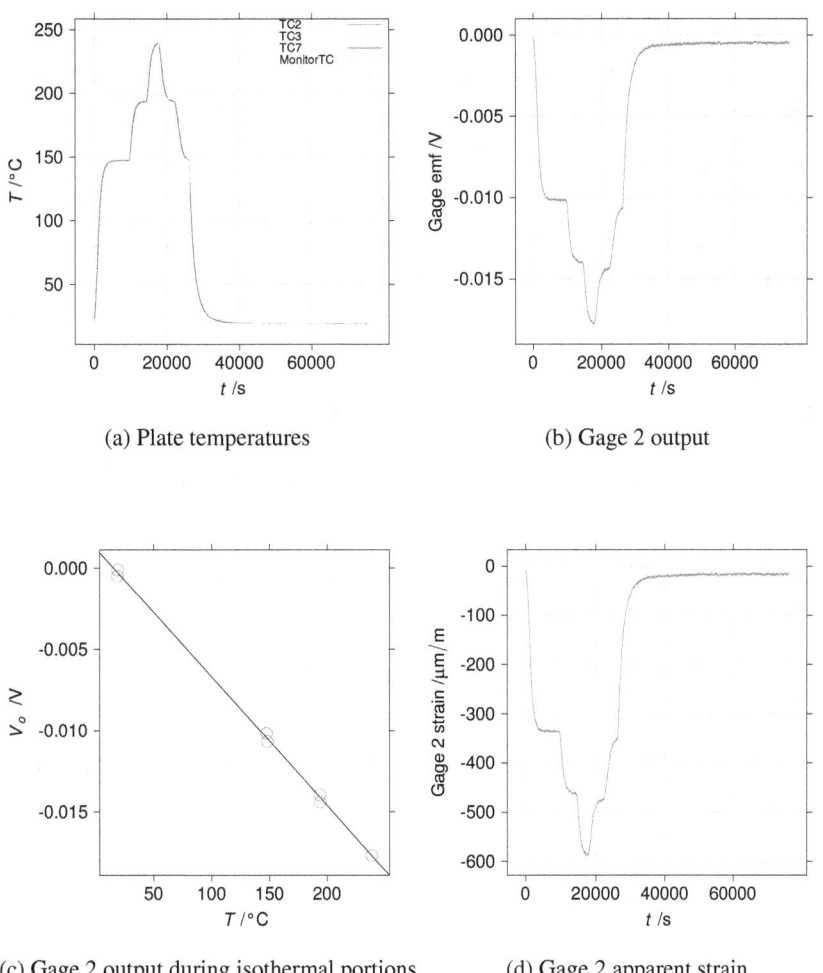

(a) Plate temperatures

(b) Gage 2 output

(c) Gage 2 output during isothermal portions

(d) Gage 2 apparent strain

Figure 24: Summary of oven test to determine the response of the gages to temperature.

compensating elements.

Apparent strains with heating in compensated and uncompensated high-temperature gages have been commonly reported [10, 11, 14]. Under the heading "Precautions for static strain measurements," the "Strain Gage Users' Handbook" states [14],

> In order to achieve accurate strain measurements at high temperature, the investigator must pre-calibrate the bonded strain gage over the test temperature range and record a zero-stress apparent strain curve.

Pre-calibrating the actual gage used in a high-temperature fire test will be very difficult, since the structure cannot be isothermally heated in a furnace. Some calibration curve could be made from a sample taken from the structural steel, but the individual gages may not be completely identical.

A.5 Findings, Conclusions, and Recommendations

Although the temperature of the plate did not reach the temperatures that steel connections reach in fire, the results of the demonstration test encouragingly support the feasibility of digital image correlation for high-temperature strain measurement in fire. Specific, significant findings include

- Commercial high-temperature paint can survive up to 700 °C for short times. (Section A.2.2)

- Rigid body translations produced no apparent strain for a plate behind a fire. (Section A.3.5)

- Digital image correlation successfully measured the thermal expansion strain of a plate heated by fire. (Section A.3.5)

- Conventional high-temperature strain gages were not fully temperature compensated, and produced apparent compressive strains on heating. (Section A.3.5 and Figure 19)

These four findings lead to four conclusions about the measurement of strains in fire conditions.

1. Commercial high-temperature paints can survive fire exposure long enough to be used in digital image correlation measurements in fire.

2. The data acquisition and analysis of the images employed no esoteric measures. The test did not employ any out-of-the ordinary illumination techniques or special image averaging, filtering, or transformations. Both of

these approaches have significant potential to improve the quality of the measurements.

3. Digital image correlation has significant promise for measuring strains and deformations in fire, and could revolutionize structural fire measurements.

4. Despite decades of development, high-temperature strain gages are still difficult to use, and have poorly understood thermal response.

The conclusions of this demonstration test lead to four recommendations for developing digital image correlation as a strain-measurement tool for the National Fire Research Laboratory.

1. A follow-on test that employs a fire powerful enough to heat the plate to more than 600 °C should be conducted. This test will probe the adherence of the paint.

2. Further tests should explore methods for post-processing the digital image correlation data to reduce the uncertainty in strains.

3. Further tests should add actual mechanical deformation deformation to the thermal expansion measured in the initial test.

4. Laboratory-scale tests in controlled environments should be used to develop strain-measurement methods for deformation so that quantifiable uncertainties can be established for the method.

References

[1] T. Chu, W. Ranson, and M. Sutton. Applications of digital-image-correlation techniques to experimental mechanics. *Experimental Mechanics*, 25:232–244, 1985. doi:10.1007/BF02325092.

[2] J Liu, J Lyons, M Sutton, and A Reynolds. Experimental characterization of crack tip deformation fields in alloy 718 at high temperatures. *Journal Of Engineering Materials And Technology-Transactions Of The ASME*, 120(1):71–78, JAN 1998. doi:{10.1115/1.2806840}.

[3] J. Lyons, J. Liu, and M. Sutton. High-temperature deformation measurements using digital-image correlation. *Experimental Mechanics*, 36:64–70, 1996. doi:10.1007/BF02328699.

[4] B. M. B. Grant, H. J. Stone, P. J. Withers, and M. Preuss. High-temperature strain field measurement using digital image correlation. *Journal of Strain Analysis for Engineering Design*, 44(4):263–271, MAY 2009. `doi:{10.1243/03093247JSA478}`.

[5] Bing Pan, Dafang Wu, Zhaoyang Wang, and Yong Xia. High-temperature digital image correlation method for full-field deformation measurement at 1200 degrees C. *Measurement Science & Technology*, 22(1), JAN 2011. `doi:{10.1088/0957-0233/22/1/015701}`.

[6] Mark D. Novak and Frank W. Zok. High-temperature materials testing with full-field strain measurement: Experimental design and practice. *Review of Scientific Instruments*, 82(11):115101, 2011. `doi:10.1063/1.3657835`.

[7] M. De Strycker, L. Schueremans, W. Van Paepegem, and D. Debruyne. Measuring the thermal expansion coefficient of tubular steel specimens with digital image correlation techniques. *Optics and Lasers in Engineering*, 48(10):978–986, 2010. `doi:10.1016/j.optlaseng.2010.05.008`.

[8] Michael A. Sutton, Jean-Jose Orteu, and Hubert W. Schreier. *Image Correlation for Shape, Motion and Deformation Measurements*. Springer, 2009. `doi:10.10007/978-0-387-78747-3`.

[9] Stephen Wnuk. Progress in high-temperature and radiation-resistant strain-gage development. *Experimental Mechanics*, 5:27A–33A, 1965. `doi:10.1007/BF02324050`.

[10] Joseph Gibbs. Two types of high-temperature weldable strain gages: NiCr half-bridge filaments and PtW half-bridge filaments. *Experimental Mechanics*, 7:19A–26A, 1967. `doi:10.1007/BF02326243`.

[11] William Sharpe. Strain gages for long-term high-temperature strain measurement. *Experimental Mechanics*, 15:482–488, 1975. `doi:10.1007/BF02318364`.

[12] ASTM International. Standard guide for high-temperature static strain measurement. Technical Report E1319-98(2009), ASTM International, W. Conshohocken, Pa, 2009. `doi:10.1520/E1319-98R09`.

[13] Y. S. Touloukian, R. K. Kirby, R. E. Taylor, and P. D. Desai. *Thermal Expansion: Metallic Elements and Alloys*, volume 12 of *Thermophysical Properties of Matter*. IFI/Plenum, New York, 1975.

[14] Steve P. Wnuk, Jr. High temperature strain gage alloys. In R. L. Hannah and S. E. Reed, editors, *Strain Gage Users' Handbook*, chapter 6, pages 191–196. Society for Experimental Mechanics, Bethel, Ct, 1992.